Exploring 3D Modeling with CINEMA 4D R19

A Beginner's Guide

Pradeep Mamgain

Exploring 3D Modeling with CINEMA 4D R19: A Beginner's Guide

NOTICE TO THE READER

Examination Copies

Textbooks received as examination copies in any form such as paperback and eBook are for review only and may not be made available for the use of the student. These files may not be transferred to any other party. Resale of examination copies is prohibited.

Electronic Files

The electronic file/eBook in any form of this textbook is licensed to the original user only and may not be transferred to any other party.

Disclaimer

No patent liability is assumed with respect to the use of information contained herein. Although every precaution has been taken in the preparation of this book, neither the author, nor PADEXI, and its dealers and distributors will be held liable for any damages caused or alleged to be caused directly or indirectly by this book. All terms mentioned in this book that are known to be trademarks or service marks have been appropriately capitalized. PADEXI cannot attest to the accuracy of this information. Use of a term in this book should not be regarded as affecting the validity of any trademark or service mark.

Book Code: PDX004P

ISBN-13: 978-1718701939

ISBN-10: 1718701934

For information about all PADEXI publications, visit our website: **www.padexi.academy**

Contents

Acknowledgments

I would like to express my gratitude to the many people who saw me through this book; to all those who provided support, offered comments, and assisted in the editing, proofreading, and design.

Thanks to:

Parents, family, and friends.

Teachers and mentors: Thank you for your wisdom and whip-cracking--they have helped me immensely.

I am grateful to my many students at the organizations where I've taught. Many of them taught me things I did not know about computer graphics.

Everyone at MAXON [www.maxon.net].

Finally, thank you for picking up the book.

This page is intentionally left blank

About the Author

I'll keep this short, I am a digital artist, writer, coder, teacher, consultant, and founder of Padexi Academy [www.padexi.academy]. I am self-taught in computer graphics, Internet has been the best source of training for me [thanks to those amazing artists, who share the knowledge for free on YouTube]. I have worked with several companies dealing with animation and VFX. I love helping young aspiring 3D artists to become professional 3D artists. I helped my students to achieve rewarding careers in 3D animation and visual effects industry.

I have more than ten years of experience in CGI. I am passionate about computer graphics that helped me building skills in particles, fluids, cloth, RBD, pyrotechnics simulations, and post-production techniques. The core software applications that I use are: Maya, 3ds Max, CINEMA 4D, Photoshop, Nuke, and Fusion. In addition to the computer graphics, I have keen interest in web design/development, digital marketing, and search engine optimization. You can contact me by sending an e-mail to **pradeepmamgain@gmail.com.**

This page is intentionally left blank

Introduction

The **Exploring 3D Modeling with CINEMA 4D R19 - A Beginner's Guide** textbook walks you through every step of creating 3D models with CINEMA 4D R19. This guide is perfect for both novices and those moving from other software to CINEMA 4D. This book will help you to get started with modeling in CINEMA 4D, you will learn important concepts and techniques about 3D modeling which you can utilize to create hard-surfaced objects for your projects. This book shares tips, tricks, notes, and cautions throughout, that will help you become a better 3D modeler and you will be able to speed up your workflow.

The first page of the every chapter summarizes the topics that will be covered in the chapter. Every chapter of this textbook contains tutorials which instruct users how things can be done in CINEMA 4D step-by-step. Practicing is one of the best ways to improve skills. Each chapter of this textbook ends with some practice activities which you are highly encouraged to complete and gain confidence for the real-world projects. By completing these activities, you will be able to master the powerful capabilities of CINEMA 4D.

Although, this book is designed for beginners, it is aimed to be a solid teaching resource for 3D modeling. It avoids any jargon and explains concepts and techniques in an easy-to-understand manner.

By the time you're done, you'll be ready to create hard-surfaced models for your 3D projects. The rich companion website PADEXI Academy (**www.padexi.academy**) contains additional CINEMA 4D resources that will help you quickly master CINEMA 4D.

What are the key features of the book?

- Learn CINEMA 4D's updated user interface, navigation, tools, functions, and commands.
- Polygon, subdivision, and spline modeling techniques covered.
- Detailed coverage of tools and features.
- Contains **24** standalone tutorials.
- Contains **14** practice activities to test the knowledge gained.
- Additional guidance is provided in form of tips, notes, and cautions.
- Important terms are in bold face so that you never miss them.

- The content under **"What just happened?"** heading explains the working of the instructions.
- The content under **"What next?"** heading tells you about the procedure you will follow after completing a step(s).
- Includes an ePub file that contains the color images of the screenshots/illustrations used in the textbook. These color images will help you in the learning process. This ePub file is included with the resources.
- Tech support from the author.
- Access to each tutorial's initial and final states along with the resources used in the tutorials.
- Quiz to assess the knowledge.
- Bonus tutorials.

Who this book is for?

This book is designed for beginners.

Prerequisites

Before jumping into the lessons of this book, make sure you have working knowledge of your computer and its operating system. Also, make sure that your that you have installed the required software and hardware. You need to install CINEMA 4D R19 on your system. Most of the tutorials will work in R17 and R18 as well.

Windows vs. Mac OS

This book is written using the **Windows** version of the CINEMA 4D. In most cases, CINEMA 4D performs identically on both Windows and Mac OS. Minor differences exist between the two versions such as difference in keyboard shortcuts, how dialog boxes/windows are displayed, and how buttons are named.

How This Book Is Structured?

This book is divided into following chapters:

Chapter M1: Introduction to CINEMA 4D R19, introduces you to CINEMA 4D interface and primitive objects available in the **Object** command group. You will learn about CINEMA 4D unit system, coordinate system, interface elements, and how to customize the interface. You will also create models using primitives.

Chapter M2: Tools of the Trade, walks you through some of the important tools that you will use in the modeling process. These tools are used to create guides in the editor view, interactively placing lights and adjusting their attributes in the scene, measure angles and distances, arrange/duplicate/randomize objects, correct lens distortions, and create virtual walkthroughs.

Chapter M3: Spline Modeling, introduces you to the spline modeling tools, concept, and techniques.

Chapter M4: Polygon Modeling, introduces you to the polygon modeling tools, concepts, and techniques. This chapter talks about polygons components, selection tools, polygons structure tools, modeling objects, and deformers.

Chapter M5: Bonus Tutorials, contains bonus tutorials.

Conventions

Icons Used in This Book

Icon	Description
	Tip: A tip tells you about an alternate method for a procedure. It also show a shortcut, a workaround, or some other kind of helpful information.
	Note: This icon draws your attention to a specific point(s) that you may want to commit to the memory.
	Caution: Pay particular attention when you see the caution icon in the book. It tells you about possible side-effects you might encounter when following a particular procedure.
	What just happened?: This icons draws your attention to working of instructions in a tutorial.
	What next?: This icons tells you about the procedure you will follow after completing a step(s).

Given below are some examples with these icons:

Tip: Framing Geometry/Elements
*You can hold down the **Alt** key with **S**, **O**, and **H** keys to zoom all views instead of the active view only.*

Note: Boolean Operations
You need to make sure the objects to which you want to apply this function should have closed volume and cleanly structured otherwise unwanted results may occur. Also, note that the higher the subdivisions of the objects, cleaner the cut will be.

Caution: Deformer Object
*The **Deformer** object does not work in conjunction with the following functions in CINEMA 4D: **Explosion Object**, **ExplosionFX**, **Polygon Reduction**, **Spline Deformer**; and **Shatter Object**.*

What just happened?
*By positioning the points, we have ensured that the both points are in a line in 3D space. It will ensure that there will be no hole in the geometry when we will apply the **Lathe** generator in the next step.*

What next?
*Now, we will create few more strips and then we will use the **Bend** deformer to create the shape of the cylinder.*

Important Words

Important words such as menu name, tools' name, name of the dialogs/windows, button names, and so on are in bold face. For example:

Select the center point. Press **MS** to invoke the **Bevel** tool and then bevel the selected edges. On the **Attribute Editor | Bevel | Tool Option** tab, change **Offset** to **22**.

Chapter Numbers

Following terminology is used for the chapter numbers and appendix:

Chapter M1, M2.......: M stands for Modeling.
Chapter MBT: MBT stands for Modeling Bonus Tutorials
Appendix AM: AM stands for Appendix A, Modeling

This approach helps us better organize the chapters when multiple modules are included in a textbook. For example, texturing chapters will be numbered as **T1**, **T2**, **T3**, and so on; the lighting chapters will be numbered as **L1**, **L2**, and so on.

Figure Numbers

In theory, figure numbers are in the following sequence **Fig. 1**, **Fig. 2**, and so on. In tutorials, the sequence is as follows: **Fig. T1**, **Fig. T2**, and so on. In tutorials, the sequence restarts from number **T1** for each tutorial.

LMB, MMB, and RMB

These acronyms stand for left mouse button, middle mouse button, and right mouse button.

MEV Menu

This acronym stands for Menu in editor view. It represents the menubar located in each viewport of CINEMA 4D.

Tool

If you click an item in a palette, toolbar, manager, or browser and a command is invoked to create/edit an object or perform some action then that item is termed as tool. For example: **Move** tool, **Rotate** tool, **Loop Selection** tool.

Right-click Popup Menus

The right-click popup menus [see Fig. 1] are the contextual menus in CINEMA 4D that provide quick access to the commands/functions/tools related to the currently selected entities.

Undo (Action)		Shift+Z
Frame Selected Elements		Alt+S, S
Create Point		M~A
Bridge		M~B, B
Brush		M~C
Close Polygon Hole		M~D
Connect Points/Edges		M~M
Polygon Pen		M~E
Dissolve		M~N
Iron		M~G
Line Cut		K~K, M~K
Plane Cut		K~J, M~J
Loop/Path Cut		K~L, M~L
Magnet		M~I
Mirror		M~H
Set Point Value		M~U
Slide		M~O
Stitch and Sew		M~P
Weld		M~Q
Bevel		M~S
Extrude		M~T, D
Array		
Clone		
Disconnect...		U~D, U~Shift+D ⚙
Melt		U~Z
Optimize...		U~O, U~Shift+O ⚙
Split		U~P

Hidden Menus

There are several hidden menus available in CINEMA 4D. These menus quickly allow you to select tools, command, and functions. For example, the **M** menu lets you quickly access the modeling tools. Now, if you want to invoke the **Extrude** tool, press **MT** [see Fig. 2].

```
Keys: M
# ... Normal Scale
, ... Normal Rotate
A ... Create Point
B ... Bridge
C ... Brush
D ... Close Polygon Hole
E ... Polygon Pen
F ... Edge Cut
G ... Iron
H ... Mirror
I ... Magnet
J ... Plane Cut
K ... Line Cut
L ... Loop/Path Cut
M ... Connect Points/Edges
N ... Dissolve
O ... Slide
P ... Stitch and Sew
Q ... Weld
R ... Weight Subdivision Surface
S ... Bevel
T ... Extrude
U ... Set Point Value
V ... Spin Edge
W ... Extrude Inner
X ... Matrix Extrude
Y ... Smooth Shift
  ... Normal Move
```

Check Box

A small box [labelled as 1 in Fig. 3] that, when selected by the user, shows that a particular feature has been enabled or a particular option chosen.

Drop-down

A drop-down (abbreviated drop-down list; also known as a drop-down menu, drop menu, pull-down list, picklist) is a graphical control element [labelled as 2 in Fig. 3], similar to a list box, that allows the user to choose one value from a list.

Button

The term button (sometimes known as a command button or push button) refers to any graphical control element [labelled as 3 in Fig. 3] that provides the user a simple way to trigger an event, like searching for a query, or to interact with dialog boxes, like confirming an action.

Window

A window [labelled as 4 in Fig. 3] is a separate viewing area on a computer display screen in a system that allows multiple viewing areas as part of a graphical user interface (GUI).

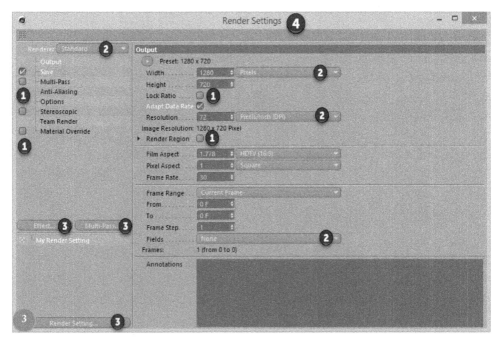

Dialog Box or Dialog

An area on screen [see Fig. 4] in which the user is prompted to provide information or select commands.

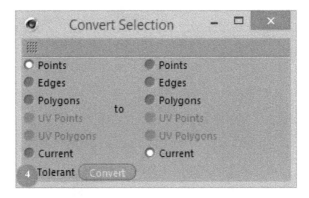

Trademarks

Windows is the registered trademarks of **Microsoft Inc. CINEMA 4D** is the registered trademarks of **MAXON Computer**.

Access to Electronic Files

This book is sold via multiple sales channels. If you don't have access to the resources used in this book, you can place a request for the resources by visiting the following link: *http://www.padexi.academy/contact*. Fill the form under the **Book Resources [Electronic Files]** section and submit your request.

Customer Support

At **PADEXI Academy,** our technical team is always ready to take care of your technical queries. If you are facing any problem with the technical aspect of the textbook, navigate to *http://www.padexi.academy/support* and open your support ticket.

Errata

We have made every effort to ensure the accuracy of this book and its companion content. If you find any error, please report it to us so that we can improve the quality of the book. If you find any errata, please report them by visiting the following link: *http://www.padexi.academy/errata*.

This will help the other readers from frustration. Once your errata is verified, it will appear in the errata section of the book's online page.

THIS CHAPTER COVERS:

- Navigating the workspace
- Customizing the interface
- Understanding UI components
- Setting preferences
- Understanding layouts
- Moving, rotating, and scaling objects
- Managers and Browsers
- Getting help

Chapter M1: Introduction to CINEMA 4D R19

Welcome to the latest version of CINEMA 4D. In any 3D computer graphics application, the first thing you see is interface. Interface is where you view and work with your scene. The CINEMA 4D's interface is intuitive and highly customizable. You can make changes to the interface and then save multiple interface settings using the **Layout** feature. You can create multiple layouts and switch between them easily.

CINEMA 4D Interface Elements

You can start CINEMA 4D by using one of the following methods:

- Double-click on the CINEMA 4D icon on the desktop
- Double-click on a CINEMA 4D scene file
- Clicking CINEMA 4D entry from the **Start** menu
- Dragging a **.c4d** file from **Windows Explorer** to the program icon
- Running it from **Command Prompt**
- Running **CINEMA 4D Lite** from the After Effects

When you first time open CINEMA 4D, you are presented with the UI, as shown in Fig. 1. I have labeled various elements of the interface using numbers. The following table summarizes the main UI elements.

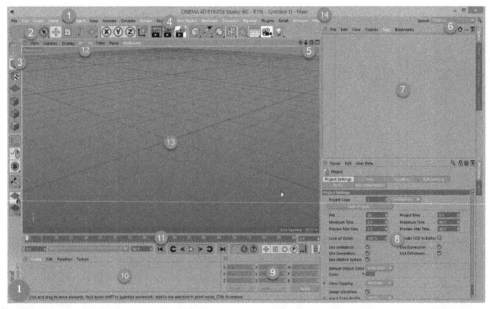

Table 1: The CINEMA 4D Interface Elements		
No.	Name	Description
1	Menubar	The menubar contains the CINEMA 4D commands and functions.
2	Standard Palette	It is located below the menubar. This palette hosts tools, commands, and functions.
3	Tools Palette	This palette contains most commonly used commands and tools.
4	Command Groups	The command groups are the collection of tools and commands.
5	Navigation Buttons	You can use these buttons to navigate in a scene.

No.	Name	Description
Table 1: The CINEMA 4D Interface Elements		
6	Layout drop-down	The options in this drop-down are used to switch between layouts.
7	Object Manager	This manager contains all objects present in the scene.
8	Attribute Manager	It displays the properties and settings of the selected objects.
9	Coordinate Manager	It contains fields that you can use to precisely model and manipulate an object in the scene.
10	Material Manager	It contains all the materials available in the scene.
11	Animation Toolbar	This toolbar contains controls for recording and playing animation.
12	Menu in editor view	This menu is available in every viewport and used to set various tasks related to the viewports. In later chapters, I've referred to it as **MEV** menubar.
13	View Panel	The view panel is collection of upto four viewports where you build and animate 3D models.
14	Title Bar	The tile bar of the main window.

The CINEMA 4D interface is highly configurable. You can dock and undock windows in the main window. When you move a docked window, the surrounding windows are resized automatically. You can also display windows as tabs to save the screen real-state. You can define the arrangements of elements as layouts and freely switch between them. The fastest way to switch between the layouts is the **Layout** drop-down [labeled as 6 in Fig. 1] located on the top-right corner of the interface. **See Video: m1vid-01.mp4**.

Let's explore various components of the CINEMA 4D interface.

Title Bar

The title bar is the first element of the interface and located at the top of the UI. It displays name and version of the software as well as the name of the currently opened file.

Menubar

The menubar is located below the tile bar. It hosts almost all important commands, tools, and functions that CINEMA 4D offers. You can tear-off a menu [or sub-menu] from the menubar by clicking on the corrugated line located at the top of the menu or sub-menu [see Fig. 2].

Standard Palette

The **Standard** palette is located below the menubar and it hosts various tools, commands, and groups of commands. Some of the icons on the palette have a black triangle on the bottom-right corner that indicates a folded group of tools/commands. To access the folded group of commands, hold LMB on the icon and then choose the desired command from the displayed flyout. The following table summarizes the tools available in the **Standard** palette:

Name	Icon	Shortcut	Description
Undo		Ctrl+Z	This tool undoes the last change. It restores scene to the previous state. By default, you can restore up to **30** previous states. If you want to change the undo depth, choose **Preferences** from the **Edit** menu to open the **Preferences** window or press **Ctrl+E**. Choose **Memory** from the list of categories and then set a new value for the **Undo Depth** attribute in the **Project** section.
1		Ctrl+Y	This tool redoes a change. It restores the changes you made in the scene.
Live Selection		9	It is like a paint brush type of selection tool. To change the brush size, change the value of the **Radius** attribute in **Attribute Editor**. To interactively change the radius, hold MMB and drag. To select entities, paint over the objects' points, edges, or polygons. You can quickly select the elements when the **Rotate**, **Move**, or **Scale** tool is active by RMB dragging over the objects. You can press **Spacebar** to toggle between the **Live Selection** tool and previously selected tool.

Table 2: The tools and commands available in the **Standard** palette

Table 2: The tools and commands available in the **Standard** palette

Name	Icon	Shortcut	Description
Rectangle Selection		0	This tool allows you to select the objects by dragging a selection frame [like marquee selection in Photoshop] over the elements. To add an element to the selection, hold **Shift** while you select. To deselect an element, hold **Ctrl** while you select.
Lasso Selection		8	This tool behaves like a lasso. You can use it to draw a loop around elements to select them.
Polygon Selection		–	This tool allows you to draw a n-sided shape to frame the elements to select them. To complete the loop, click on the starting point or RMB click. **See Video: m1vid-02.mp4**.
Move		E/4/7	Allows you to place the selected object or component in the viewport. You can also use it to select points, edges, and polygons. The selection can also be drawn using RMB. You can press the **4** key and drag the mouse to move the objects. Press **7** and drag the mouse to move the objects without child objects. You can also use the **Axis Extension** feature to accurately position the objects. You need to **Ctrl+RMB** click on an axis handle to enable the **Axis Extension** feature. **See Video: m1vid-03.mp4**.
Scale		T	Allows you to resize the selected objects or components. The **Axis Extension** function is also available for the **Scale** tool.
Rotate		R	Allows you to rotate the selected objects or components.

Table 2: The tools and commands available in the **Standard** palette

Name	Icon	Shortcut	Description
Active	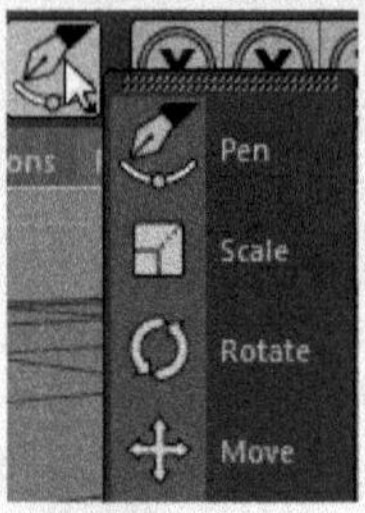		A list of last used 8 tools is displayed here when you click and hold on it.
X-Axis / Heading		X	This tool locks or unlocks the transformation along the X-axis.
Y-Axis / Pitch		Y	This tool locks or unlocks the transformation along the Y-axis.
Z-Axis / Bank		Z	This tool locks or unlocks the transformation along the Z-axis.
Coordinate System		W	This tool is a toggle switch that allows you to switch between the local and world coordinate systems.
Render View		Ctrl+R	Renders the current active view.
Render to Picture Viewer		Shift+R	Renders the active view in the **Picture Viewer** window.
Edit Render Settings		Ctrl+B	Opens the **Render Settings** window from where you can specify settings for rendering the scene.
Cube			Click on the **Cube** tool to create a cube. Hold **LMB** on this tool to reveal the **Objects** command group. This group holds the geometric primitives that CINEMA 4D offers.
Pen			The **Pen** tool is a versatile tool that is used to create and edit splines. Hold **LMB** on this tool to reveal the **Spline** command group. This group holds the spline primitives and other spline tools CINEMA 4D offers.

Table 2: The tools and commands available in the **Standard** palette

Name	Icon	Shortcut	Description
Subdivision Surface			This tool is a generator that allows you to easily create smooth 3D models with low poly count. This tool supports point weighting, edge weighting, and so on. Hold **LMB** on this tool to reveal the **Generators** command group. This group holds the generators that CINEMA 4D offers.
Array			This tool creates copies of the objects and arranges them in a spherical or wave form. Press and hold LMB on this tool to reveal the **Modeling** command group. This group holds various modeling commands.
Bend			This tool bends the objects. Hold **LMB** on this tool to reveal the **Deformer** command group. This group holds the deformer functions CINEMA 4D offers.
Floor			This tool creates a floor object that always lies in the XZ plane of the world coordinate system. It stretches to infinity in all directions. Hold **LMB** on this tool to reveal other environment tools available in the **Environment** command group.
Camera			This tool creates a camera in the scene. Hold **LMB** on this tool to reveal other camera tools available in the **Camera** command group.
Light			This tool creates a light in the scene. Hold **LMB** on this tool to reveal other light tools available in the **Light** command group.

Note: Highlighting and Selections

The selected objects are surrounded by an orange outline. If you hover the mouse over objects in the editor view, a white outline appears on the objects underneath the mouse pointer. If you are using a graphics card that is not supported by CINEMA 4D, these highlights will not appear. In case you have a supported graphics card and highlights are not appearing, you need to *enable the **Enhanced OpenGL** option. You can access this option from the **Options** menu of the **MEV** menubar.*

Tools Palette

The **Tools** palette is the vertical toolbar located on the extreme left of the standard interface. It contains various tools. The following table summarizes these tools:

Name	Icon	Shortcut	Description
Make Editable		C	The primitive objects in CINEMA 4D are parametric. They have no points or polygons and are instead created using math. This tool allows you to convert the parametric objects to editable objects with polygons and points. When you make an object editable, you lose its parametric creation parameters.
Model			If you want to move, rotate, or scale an object, use this mode.
Object			This icon is available when you press and hold the left mouse button on the **Model** icon. The **Object** mode is suited when you are working with an animation. When you scale an object using the **Scale** tool, only the object axes are scaled not the surfaces themselves. If you scale an object non-uniformly, the child objects get squashed and stretched when you rotate them. You can avoid this problem by working in the **Object** mode. The rule of thumb is when modeling, use the **Model** mode, use the **Object** mode for animation.
Animation			This icon is available when you press and hold the left mouse button on the **Model** icon. This mode allows you to move, scale, or rotate the entire animation path of an object.
Texture			This mode allows you to edit the active texture. Only one texture can be edited at a time. Also, note that the projection type is also taken into consideration in this mode.
Workplane			Workplanes are explained in detail in Chapter M2.

Table 3: The tools and commands available in the **Tools** palette

Table 3: The tools and commands available in the **Tools** palette

Name	Icon	Shortcut	Description
Points			Select this tool to enable the **Points** mode. Once enabled, you can use various tools to edit the points of an object.
Edges			Click this tool to enable the **Edges** mode and edit the edges of an object.
Polygons			Click this tool to enable the **Polygons** mode and edit the polygons of an object.
Enable Axis		L	This tool allows you to move the origin of the object.
Tweak			This tool works with the **Polygon Pen** tool. You can tweak points, edges, and polygons using the **Polygon Pen** tool in this mode.
Viewport Solo Mode			This mode is useful when you are working on a complex scene and you want to concentrate on a specific area of the scene. The **Viewport Solo Off** option is a toggle switch that you can use to turn on or off the **Viewport Solo** mode. In the **Viewport Solo Single** mode, all objects are hidden except the currently selected objects. The **Viewport Solo Hierarchy** mode displays only the selected objects, including their child objects. When the **Viewport Selection Solo** option is selected, the visibility of the objects will be defined automatically when they are selected.
Enable Snap		Shift+S	It is a toggle switch that lets you enable or disable snapping.
Locked Workplane		Shift+X	This tool disables any defined automatic **Workplane** modes and fixes the **Workplane**.
Planar Workplane			Depending on the angle of view of the camera, one of the world coordinate planes will be automatically displayed as the **Workplane**.

Navigation Tools

The navigation tools are located on the top-right corner of each viewport. Click-drag the first icon to pan the view [move the camera], click-drag second icon to zoom in or out [move the camera in the direction of view] of the viewport. Click-drag the third icon to rotate [rotate the camera] the view. Click the forth icon to maximize the view. You can also maximize a viewport by **MMB** clicking on it. You can also use hotkeys to navigate the view. To use a hotkey, hold down the key on the keyboard and drag the mouse. The following table summarizes these hotkeys:

Table 4: The navigation hotkeys	
Key	**Description**
1	Move camera
2	Move camera in the direction of view
3	Rotate Camera

The viewports with the orthogonal view can be rotated around their orthographic axis. If you hold **Shift,** you can rotate the camera in **15** degrees increments.

Menu In Editor View

The **Menu in editor view** [MEV menubar] is located on the top of each viewport. It hosts command, tools, and functions corresponding to the views. The **MEV** menubar contains seven menus. The following tables summarizes the options available in these menus:

Table 5: The **View** menu			
Option	**Icon**	**Shortcut**	**Description**
Use as Render View			When enabled, the active camera will be used for the rendering in **Picture Viewer**.
Undo View/ Redo View		Ctrl+Shift+Z Ctrl+Shift+Y	These tools only work in the viewports. The main **Undo/ Redo** tools do not affect the cameras.
Frame All			Centers all objects including lights and cameras to fill the view.
Frame Geometry		Alt+H H	Centers all objects excluding lights and cameras to fill the view.
Frame Default			Resets the viewport to the default values.
Frame Selected Elements		Alt+S S	Centers all selected elements [objects, polygons] to fill the view and are centered.

Table 5: The **View** menu

Option	Icon	Shortcut	Description
Frame Selected Objects		Alt+O O	Centers all active objects to fill the view.
Film Move			The functioning of this tools is similar to that of the first icon in the navigation tools, refer to Table 4.
Redraw		A	Redraws the scene. By default, CINEMA 4D updates the scene automatically. If it does not happen, use this function to redraw the scene.

Tip: Framing Geometry/Elements
*You can hold down the **Alt** key with **S**, **O**, and **H** keys to zoom all views instead of the active view only.*

Table 6: The **Cameras** menu

Option	Icon	Description
Cursor Mode		Every viewport has a default camera called **Editor Camera**. This camera is active by default. These option available in the **Navigation** sub-menu define how camera is rotated. If you have defined a camera and using it, the camera will be always rotated around its origin independent of **Camera Mode**. The **Cursor** mode is the default mode. In this mode the camera rotates around the selected object point. If you click on an empty area of the viewport, the camera will pivot around that point.
Center Mode		In this mode, the camera will rotate around the center of the screen.
Object Mode		In this mode, the camera will rotate around the center of the selected objects/elements.
Camera Mode		In this mode, the camera will rotate around its own axis.
Default Camera		The command activates the default camera.

Table 6: The **Cameras** menu

Option	Icon	Description
Set Active Object as Camera		You can use this function to view the scene from the origin of the active object.
Perspective		This is the default projection mode for the viewport [see Fig. 3]. You see the scene as if you are looking through a conventional camera.
Parallel		All lines are parallel. The vanishing point is infinitely distant [see Fig. 4].
Left		The YZ view [see Fig. 5].
Right		The ZY view [see Fig. 6].
Front		The XY view [see Fig. 7].
Back		The YX view [see Fig. 8].
Top		The XZ view [see Fig. 9].
Bottom		The ZX view [see Fig. 10].
Axonometric		There are six more **Axonometric** views available. The **Axonometric** projection is a type of parallel projection used for creating a pictorial drawing of an object, where the object is rotated along one or more of its axes relative to the plane of projection.

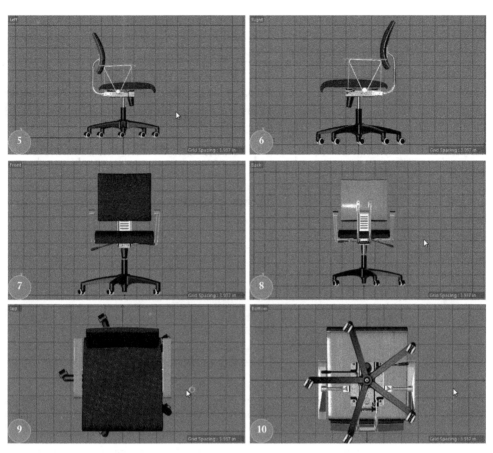

Table 7: The **Display** menu

Option	Icon	Shortcut	Description
Gouraud Shading		N~A	This is the high quality display mode for viewports [see Fig. 11]. The object smoothing and lights are taken into consideration.
Gouraud Shading (Lines)		N~B	Adds wireframes or isoparms to the shading [see Fig. 12].

Quick Shading		N~C	Identical to **Gouraud Shading**, however, the auto light is used instead of scene's lights.

Table 7: The **Display** menu

Option	Icon	Shortcut	Description
Quick Shading (Lines)		N~D	In this mode, you can add wireframes or isoparms to the quick shading by choosing **Wireframe** or **Isoparms** from the **Display** menu.
Constant Shading		N~E	Shows constant shading on the objects.
Constant Shading (Lines)			Shows constant shading with lines on the objects.
Hidden Line		N~F	The hidden lines are not displayed.
Lines		N~G	Displays complete mesh including hidden lines.
Wireframe		N~H	Draws lines on objects.
Isoparms		N~I	This mode displays isoparm lines for objects that use them such as **Generators**.
Box		N~K	Displays each object as box.
Skeleton		N~L	This is the fastest display mode. It is only suitable for hierarchical structures.

Table 8: The **Options** menu

Option	Icon	Shortcut	Description
[Level of Detail] Low Medium High			These options define level of detail in the viewport. The **Low**, **Medium**, and **High** options set the level of detail in the viewport to **25%**, **50%**, and **100%**, respectively.
Use Render LOD for Editor Rendering			Using this option, you can define LOD detail for each view. This option lets you use the LOD settings defined in the respective settings such as subdivision surfaces or metaballs.

Table 8: The Options menu

Option	Icon	Shortcut	Description
Stereoscopic			Enables the stereoscopic display in the view.
Linear Workflow Shading			If you are considering a stereoscopic view, the color and shaders can be turned off using this option.
Enhanced OpenGL			Defines whether the viewport should use the **Enhanced OpenGL** quality for display.
Transparency			Defines whether or not **Enhanced OpenGL** should display transparency in high quality.
Shadows			Defines whether or not **Enhanced OpenGL** should display shadows.
Post Effects			Defines whether or not **Enhanced OpenGL** should display post effects.
Noises			Toggles the display of the **Noise** shader for **Enhanced OpenGL**.
Reflections			Defines whether or not **Enhanced OpenGL** should display reflections.
SSAO			Defines whether or not an **Ambient Occlusion** approximation should be rendered in the Viewport.
Tessellation			Defines whether or not tessellation should be displayed for respective objects in the Viewport.
Depth of Field			Defines whether or not the depth of field should be rendered in the Viewport.
Backface Culling		N~P	Toggles backface culling on or off when in the **Lines** mode. With backface culling, all concealed surfaces are hidden from the camera improving the performance.

Table 8: The **Options** menu

Option	Icon	Shortcut	Description
Isoline Editing		Alt+A	It projects all **Subdivision Surfaces** cage object elements onto the smoothed surface. As a result, these elements can be selected directly on the smoothed object.
Layer Color			This option lets you view which objects have been assigned to which layer. The objects are displayed in the color assigned to their respective layer.
Polygon Normals			Toggles the display of normals in the polygon mode.
Vertex Normals			Toggles the display of normals in the vertex mode.
Tags		N~O	If enabled, the objects will use the display mode defined in their **Display** tags.
Textures		N~Q	Allows you to see textures in the view panel in real-time.
X-Ray		N~R	Enables the X-Ray effect. The object becomes semi-transparent so that you can see its concealed points and faces.
DefaultLight			Opens the **Default Light** manager that you can use to quickly light the selected objects from any angle. Click-drag on the sphere to set the angle of light.

Table 8: The **Options** menu			
Option	**Icon**	**Shortcut**	**Description**
Configure		Shift+V	The **Configure** option is used to specify the viewport settings. The settings are displayed in **Attribute Manager**. The options and parameters displayed in bold face in **Attribute Manager** are saved globally and these options are used when you create a new scene or restart CINEMA 4D.

All non-bold parameters and options are saved with the file locally. The local options and parameters affect the active view or the selected view. You can select multiple views by clicking on the blank gray area of their headers with **Shift** held down. You can make an option or a parameter global or local. To do this, select the element by clicking on it in **Attribute Manager** and then RMB click. Choose **Make Parameter Global/Make Parameter Local** from the popup menu. |
| Configure All | | Alt+V | This option affects all existing views. |

The options in the **Filter** menu allow you to define which types of objects are displayed in the views. By default, all types are displayed. Choose **All** from the menu to enable all types. Choose **None** to disable all types.

 Tip: Filters
*To enable one option and disable all others, choose the desired filter from the Filter menu with **Ctrl** held down.*

 Note: Hidden Objects
*If you select a hidden object from **Object Manager**, the axis system of the object appears in the viewport.*

Each view in CINEMA 4D can have upto four view panels. The options available in the **Panel** menu allow you to choose a different mode [arrangement of viewports] for the view panel. Fig. 13 shows the viewport arrangement when I chose **Arrangement | 4 Views Top Split** from the **Panel** menu.

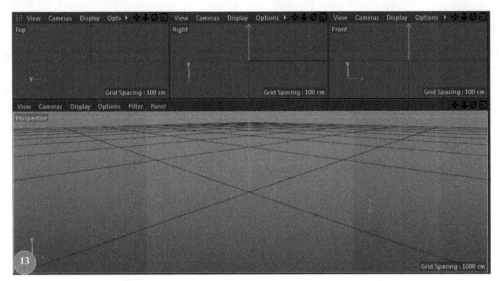

You can use **Function** keys to toggle between the full-screen and normal size. You can also access the corresponding functions from the **Panel** menu. The following table summarizes these keys:

Table 9: The **Function** keys	
Key	**Description**
View 1 (F1)	Maximizes the **Perspective** view.
View 2 (F2)	Maximizes the **Top** view.
View 3 (F3)	Maximizes the **Right** view.
View 4 (F4)	Maximizes the **Front** view.
View 5 (F5)	Restores four viewports.

Table 10: The **ProRender** menu		
Option	**Icon**	**Description**
Use as ProRender View		Use this option to set the current view as the preview view. You can select ProRender options only if you have set **Renderer** to **ProRender** in the **Render Settings** window. Any view can be defined as an interactive preview renderer. You can also navigate in the view and reduce resolutions to achieve faster results interactively. The changes that you make to materials or cameras will be immediately visible in the ProRender view. When you select the **Use as ProRender View** option, a small HUD will be displayed at the bottom of the ProRender view to start/stop the rendering or select the render settings.

Table 10: The **ProRender** menu		
Option	**Icon**	**Description**
Start ProRender		Use this option to start rendering in the preview view. The speed of rendering depends on the complexity of the scene. Select this option again to switch back to the normal OpenGL view.
Use Offline Settings		These two options are used to select the settings to be used for the ProRender view to be rendered.
Use Preview Settings		
Camera Updates		These options are used to define if the preview render should be updated if one the elements such as camera, light, material, or geometry is modified.
Material Updates		Normally, all elements are updated by default. Due to technical reasons, the rendering can be occasionally be restarted without ending while you are working on **MoGraph, Hair,** or **XPresso.** If you disable **Geometry Updates,** that will help you in preventing constant updating.
Light Updates		
Geometry Updates		
Synchronize Viewport		This option is useful when you are working in two views. For example, one ProRender view and one OpenGL view. In such cases, if you move the camera in the OpenGL view, the camera movement will be carried over to the ProRender view.
Send to Picture Viewer		You can use this option to load the image rendered by the ProRender view to the **Picture Viewer** window.

HUD

The term HUD is taken from the aviation industry. In aircraft, HUD refers to the projection of reading on a screen so that the pilot can read the values without looking down. HUD in CINEMA 4D does the same function. The HUD can be switched on and off from the **Filter** tab of the **Viewport** settings. To access this tab, choose **Configure** from the **MEV's Options** menu. Now, in **Attribute Editor,** choose the **Filter** tab. From this tab, you can toggle HUD using the **HUD** check box. Now, choose the **HUD** tab and select the checkboxes as per the requirement. For example, if you want to see HUD for the number of polygons in the scene, select the **Total Polygons** check box. If you want to see number of the selected polygons as well, select the **Selected Polygons** check box, refer to Fig. 14.

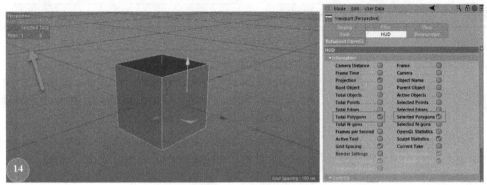

You can also add objects parameters to the HUD. To do so, select the attribute(s) in **Attribute Manager** and then RMB. Choose **Add to HUD** from the popup menu to add selected attributes to the HUD. You can also drag and drop attributes from **Attribute Manager** to editor view. Now, you can click drag an attribute to adjust its value. You can use the following keys with HUD:

- Use **Shift** to select multiple HUD elements.
- Use **Ctrl** to move HUD elements.
- Double-click on a HUD element to open a text field that allows you to enter a new value for the element.

Managers and Browsers

There are variety of managers, explorers, and browsers that CINEMA 4D offers. Given below is a quick rundown:

Object Manager

Object Manager is the nerve center for all objects and their corresponding tags in a scene. This manager allows you to manage object's name, hierarchy, visibility, and so on. It also allows you to manage tags that have been assigned to the objects. Some of the functions of **Object Manager** are given next:

- To select an object, click on it. To select multiple objects, click with **Ctrl** held down. You can use **Shift** to select an entire range of objects. You can also select multiple objects by holding down **Shift** and using the **Up** and **Down** arrow keys. The objects can also be selected by drawing a marquee selection. When you select an object, its settings are displayed in **Attribute Manger**. When you select multiple objects, common settings are displayed in **Attribute Manager**.
- **MMB** click selected an object including all its children.
- **Alt+MMB** click selects the object you click and all objects on the same hierarchy level except children.
- The selected items are color highlighted. The last selected object highlighted in slightly lighter color. Child objects of the selected objects are also highlighted in a lighter color.
- To open and close branches in **Object Manager**, click + or - sign. If you want to open or close the entire hierarchy, click + or - sign with **Ctrl** held down.
- Press the left or right arrow key to close or open the active branches.

- To make copies of an object, drag it with **Ctrl** held down.
- You can rearrange items in the manager by dragging and dropping.
- To rename items, double-click on their names. You can also select an item and press **Enter** to enable a rename field in which you can type the new name. When rename field appears, you can use the up and down arrow keys to quickly rename the items.

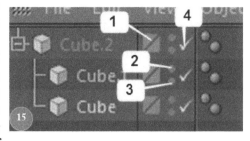

- The second column in **Object Manager** [see Fig. 15] contains some switches labeled from 1 to 4 in Fig. 14 [sometimes also referred to as **Traffic Lights**]. The **Layer color** switch, labeled as 1, displays color of the layer. Click on this switch to open a popup menu. You can add the object to a layer or open **Layer Manager** to have a greater control on layers and objects they host.
- The switch labeled as 2, is the **Editor On/Off** switch. It allows you to control the visibility of the object in the editor view. By default, the color of this switch is gray, which is the default behavior. Click on it to override the default behavior and turn the visibility of the object on. The color of the switch turns green. Click one more time to hide the object from the editor view. On doing so, the color of the switch turns red. The green dot enforces the visibility of an object. If you have grouped several objects [Hotkey: **Alt+G**] and now you turn off the visibility of the group using red dots, you can still make the objects visible from the group by using the green dots.
- The switch labeled as 3 allows you to visibility of the objects in renders.
- The switch labeled as 4 affects the object in the editor view as well as in the renderer. It essentially a way to turn off an object completely.
- The right side of the switches column is the area where you will find the tags associated with the objects.

Rearranging Objects

There are many ways for rearranging objects in **Object Manager**. When you drag the objects, different icons appears on the mouse pointer indicating what action CINEMA 4D will take once you drop them. The following table summarizes these icons.

Table 11: Rearranging objects

Icon	Description
	Drag an object between two other objects or to the end of the list.
	Ctrl+Drag to create a copy.

Table 11: Rearranging objects	
Icon	**Description**
	Makes the dragged object a child of the other.
	Use **Ctrl+drag** and move the mouse pointer over an object to create a copy and make it a child of another object.
	You can also drag-and-drop tags. To transfer a tag from one object to another, drag the tag icon on to the line of the other object.
	If you want to create a copy, use **Ctrl+Drag**.
	No operation is available.

Attribute Manager

Attribute Manager allows you to specify value for almost every parameters in CINEMA 4D. You can access parameters for objects, tools, materials, and so on from this manager. You can also animate parameters in **Attribute Manager**. By default, it displays attributes of the selected object. If you are using an object frequently, you can create a copy of **Attribute Manager** and then lock it to that object. To create a new copy, click on the + icon [ ] located on the top-right corner of the title bar. To lock the manager, click the lock icon [ ] on the manager's title bar.

Coordinate Manager

Coordinate Manager allows you to manipulate objects numerically. It displays fields for editing position, scale, and rotation values. You can also use it as a reference when you are scaling objects interactively in the editor view. The values are displayed in conjunction with the tool you are using. For example, if you are using the **Move** tool; the position, size, and rotation values of the selected element are shown in the fields.

Note: Scaling Objects
*Scale the objects only when there is no other way. For example, if you want to make a sphere bigger, use its **Radius** attribute instead of scaling it up. Also, see the **Object** mode description in Table 3.*

Material Manager

Material Manager allows you to create materials and apply them onto the objects in the scene. The thumbnail of each material you create is displayed in the manager. When you select an object in **Object Manager**, the thumbnails of the materials applied to that object appear depressed. To apply a material to an object, drag the material's thumbnail from **Material Manager** and drop it on an object(s) in **Attribute Manager** or in the editor view.

Take Manager

When you are working on a complex project that contains various animations, render settings, cameras, and so on, you will have to prepare and maintain several project files. Working on different files wastes a lot of time and effort. **Take Manager** [see Fig. 16] allows you to overcome this issue by saving multiple settings in one file and then that file can be rendered with powerful variable file and path names (**Tokens**). **Take Manager** lets you save the initial state as the master take and then you can create new animations to fine-tune and test the scene.

Content Browser

Content Browser [see Fig. 17] allows you to navigate through the content libraries that come with CINEMA 4D. You can easily import additional content and presets into your projects. You can use this manager to manage scenes, images, materials, shaders, and presets. This browser lets you manage file structure of the scene.

Structure Manager

Structure Manager [see Fig. 18] shows the data related to an object if the selected object is editable. This manager shows data like a spreadsheet. It contains cells that are divided into rows and columns. The data shown in the cells depends on the mode you are in. The following data is shown in **Structure Manager**:

- Points
- Polygons
- UVW Coordinates
- Bezier Spline Tangents
- Weight Vertices
- Normal Vector Coordinates

The values shown in the cells can be directly edited. You can also drag and drop the lines. The editing functions such as cut, copy, and paste are also supported.

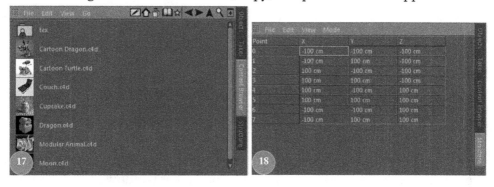

Layer Manager

The **Layer Manager** [see Fig. 19] displays all layers that you have created. It is very useful when you are working on a complex scene. It lets you manage a complex scene easily. You can assign a custom color to the layer that also appears in **Object Manager**. You can drag and drop a layer onto an object to assign that object to the dragged layer. If you hold down the **Ctrl** key while dragging and dropping a layer onto an item, the layer is assigned to the item's children as well.

Project Settings

You can use the **Project Settings** [see Fig. 20] to define standard values such as animation time, scene scale, and so on for the current scene. These are the settings that affect the scene globally. You can open **Project Settings** by choosing **Project Settings** from the **Edit** menu. Alternatively, you can press **Ctrl+D**. You can also open it by choosing **Mode | Project** from the **Attribute Manager's** menu bar.

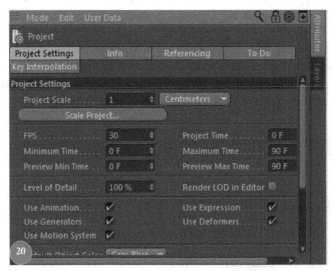

Help

The CINEMA 4D help documentation can be accessed by choosing **Help** from the **Help** menu. Like any other window in CINEMA 4D, the help window can be docked anywhere in the interface. CINEMA 4D also supports context sensitive help. If you want to access help for a button, tool, icon, and so forth, hover the mouse pointer on the element and then press **Ctrl+F1**. You can also RMB click on an attribute and then choose **Show Help** from the popup menu to see help documentation about that attribute.

If you hover the mouse pointer over almost any item in CINEMA 4D, a brief description about the item appears in the bottom-most window of the interface.

Commander

The **Commander** window in CINEMA 4D is used to call up commands, objects, tools, and tags without using any manager. You can invoke the **Commander** window by clicking on the magnifying glass icon located next to the **Layout** drop-down in the top-left corner of the interface.

Alternatively, you can press **Shift+C** to open it. Type the name of the entity you are looking for; CINEMA 4D will display a list of matching commands [see Fig. 21]. Select the desired option from the list. You can press **Esc** to close the **Commander** window.

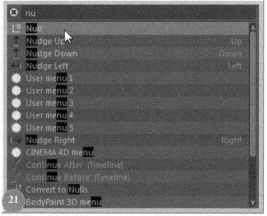

Hidden Menus

There are several hidden menus available in CINEMA 4D. These menus quickly allow you to select tools, command, and functions. The **V** menu [see Fig. 22] provides a useful shortcut to quickly switch between the view, selection, tools/modes, plugins, and snapping options.

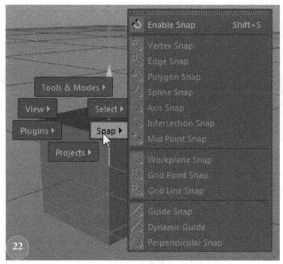

The **M** menu [see Fig. 23] lets you quickly access the modeling tools. For example, if you want to invoke the **Extrude** tool, press **M+T**. The **N** menu [see Fig. 24] clones the options from the **Display** menu of the editor's menubar. The **P** menu allows you to access snapping functions and commands [see Fig. 25].

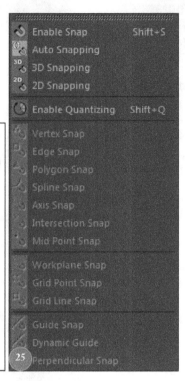

```
Keys: M
# ... Normal Scale
, ... Normal Rotate
A ... Create Point
B ... Bridge
C ... Brush
D ... Close Polygon Hole
E ... Polygon Pen
F ... Edge Cut
G ... Iron
H ... Mirror
I ... Magnet
K ... Knife
L ... Set Point Value
O ... Slide
P ... Stitch and Sew
Q ... Weld
R ... Weight Subdivision Surface
S ... Bevel
T ... Extrude
W ... Extrude Inner
X ... Matrix Extrude
Y ... Smooth Shift
23  Normal Move
```

```
Keys: N
A ... Gouraud Shading
B ... Gouraud Shading (Lines)
C ... Quick Shading
D ... Quick Shading (Lines)
E ... Constant Shading
F ... Hidden Line
G ... Lines
H ... Wireframe
I ... Isoparms
K ... Box
L ... Skeleton
O ... Tags
P ... Backface Culling
Q ... Textures
24  X-Ray
```

```
            Enable Snap          Shift+S
            Auto Snapping
  3D        3D Snapping
  2D        2D Snapping

            Enable Quantizing    Shift+Q

            Vertex Snap
            Edge Snap
            Polygon Snap
            Spline Snap
            Axis Snap
            Intersection Snap
            Mid Point Snap

            Workplane Snap
            Grid Point Snap
            Grid Line Snap

            Guide Snap
            Dynamic Guide
  25        Perpendicular Snap
```

Tutorials

Before you start the tutorials, let's first create a project folder that will host the tutorial files. Open **Windows Explorer** and create a new folder with the name **chapter-m1**.

Tutorial 1: Creating a Sofa

In this tutorial, we will model a sofa using the **Cube** primitive [see Fig. T1].

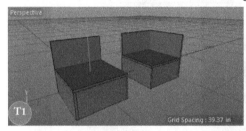
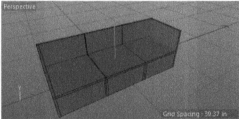

Table T1
Flow: The following sequence will be used to create the sofa model: **1.** Understand the coordinate system in CINEMA 4D. **2.** Work with CINEMA 4D units system. **3.** Create the sofa model using the **Cube** primitive.

Skill level	Beginner
Time to complete	30 Minutes

Table T1	
Topics in the section	• Getting Your Feet Wet • Creating One Seat Section of the Sofa • Creating Corner Section of the Sofa
Project Folder	**chapter-m1**
Units	**Inches**
Final tutorial file	**m1-tut1-finish.c4d**

Getting Your Feet Wet

Follow the steps given next:

1. Choose **File | New** from the main menubar or press **Ctrl+N** to start a new scene. Choose **Object | Cube** from the **Create** menu to create a cube at the center of the **Perspective** view. Ensure any selection tool is active in the **Standard** palette and then click and drag the green handle [along +Y-axis]; notice that the new **Y** position value is displayed in **Coordinate Manager** and **Status Bar** at the bottom of the UI [see Fig. T2].

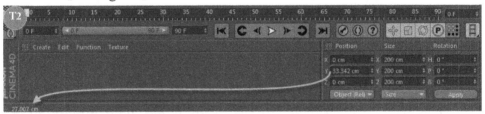

2. In the **Coordinate Manager | Position [Y]** field, enter **0** to place the cube back at the origin.

3. Choose **World** from the drop-down located below the **Position** fields in the **Coordinate Manager** and then enter **100** in the **Position [Y]** field. The cube now sits at a distance of 100 units from the origin.

4. In **Object Manager**, double-click on the text **Cube** to enable the edit field and then rename it as **myCube**. Notice the name is displayed in **Attribute Manager** as well.

5. In **Attribute Manager**, ensure **Coord** parameter group is selected [see Fig. T3] and then in **Freeze Transformation** section, click **Freeze P**.

> **?** *What just happened?*
> *Freezing transformations in CINEMA 4D freezes [zero out, also referred to as dual transformation] the coordinates of all selected objects. When you freeze coordinates, the local position and rotation coordinates will each be set to o and scale to 1 without changing position or orientation of the object.*

? *Freezing transformations is particularly useful in animations using parent-child relationships. When you rotate a child object around an axis, all three axes are affected because the parent object's coordinate system has a different orientation from the local coordinates. You can avoid this scenario by freezing the coordinates before animating.*

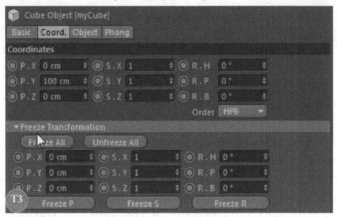

Tip: Parameter Groups/Tabs
*The parameters of an object are categorized in different groups or tabs in **Attribute Manager**. Notice in Fig. T3 there are four groups or tabs: **Basic**, **Coord**, **Object**, and **Phong**. The selected group button appears in the light blue color [**Coord** in this case]. If you want to display parameters from different groups in **Attribute Manager**. Click on the parameter buttons with **Shift** or **Ctrl** held down. You can also click-drag on the buttons to display parameters from different groups.*

6. Choose **Object (Rel)** from the first drop-down in **Coordinate Manager** and then enter **50** in the **Position [Y]** field. Notice now the cube moves **50** units up from its current position.

? **What just happened?**
*Object (Rel) defines the relative object location without frozen coordinates. This is same as the main coordinates in the **Coord** parameter group of the **Attribute Manager**. **Object (Abs)** defines the relative object location as a combination of frozen coordinates and **Object (Rel)** coordinates. **World** defines the object location in world units.*

7. In **Object Manager**, drag **myCube** with **Ctrl** held down to create a copy of the cube with the name **myCube.1**. In the editor view, drag the green handle in the positive Y-axis to display both cubes [see Fig. T4].

8. Drag **myCube.1** onto the **myCube** and release the mouse button when an icon similar to the one shown in Fig. T5 to make **myCube.1** child of **myCube** [see Fig. T6].

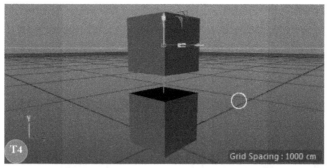

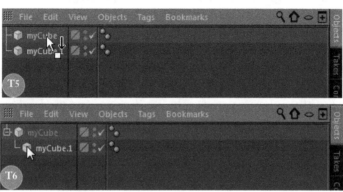

9. Select **myCube** in **Object Manager**. Choose **Size+** from the middle drop-down in **Coordinate Manager**. Notice that the size of the parent+child is displayed in the **Scale** fields.

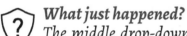
What just happened?
The middle drop-down in **Coordinate Manager** *specifies which object size is shown in the* **Scale [XYZ]** *fields.* **Size** *shows the size of the object without considering children [see Fig. T7] whereas* **Size+** *considers the children as well [see Fig. T8]. Scale shows the axis length of the object coordinate system. Default value is 1:1:1 [see Fig. T9].*

Note: Scale and Parent-child Relationship
When you create a child of a parent, the scale of the child axis adjusted according to the parent so that the child appears normal with respect to the world axes. For example, if scale of the parent object is **4:1:1,** *when you create a child for this object, the scale of the child axes will be* **0.25:1:1**.

10. Choose **Close** from the **File** menu or press **Ctrl+F4** to close the scene. Do not save the file.

Now, we understood CINEMA 4D's coordinate system therefore now let's move ahead and build the sofa. Before we jump into modeling, let's first set the units for the project.

Specifying Units

You can define units for the project from two locations in CINEMA 4D: **Preferences** window and **Project Settings**. Choose **Preferences** from the **Edit** menu or press **Ctrl+E**. Choose **Units** from the list of categories and then choose **Inches** from the **Unit Display** drop-down of **Units | Basic** section. Close the **Preferences** window.

> *What just happened?*
> *Here, I chose **Inches** as units for the project. This setting does not affect the scene parameters [scale]. It just converts the values in the fields. For example, if you have defined a value of **20 cm** and you switch units to **Meters**, the filed will display the value **0.2m**. However, if you want to change the scale of the scene, you can do so from the **Project Settings** by adjusting the **Project Scale** parameter. If you change units to meters, a **20cm** wide object will be scaled to **20m** wide object.*

> *Tip: Project Settings*
> *You can open **Project Settings** in **Attribute Manager** by choosing **Project Settings** from the **Edit** menu, or press **Ctrl+D**. To change scale of the scene, click **Scale Project** from **Attribute Manager** to open the **Scale Project** dialog. In this dialog, set the **Target Scale** and then click **OK** to change the scale. This feature is specifically useful when you are importing an object created in an external application and does not have a correct scale.*

Press **Ctrl+S**, the **Save File** dialog appears. Navigate to the **chapter-m1** folder and save the file with the name **m1-tut1-finish.c4d**.

> *Note: Saving Files*
> *I highly recommend that you save your work at regularly by pressing **Ctrl+S**.*

Creating One Seat Section of the Sofa

Follow the steps given next:

1. On the **Standard** palette, click **Cube** to create a cube in the editor view. On the **Attribute Manager | Object** tab, enter **25.591, 1,** and **25.591** in the **Size X, Size Y,** and **Size Z** fields, respectively.

2. Press **Alt+O** to frame the cube in the editor view. Press **NB** to enable **Gouraud Shading (Lines)** display mode [see Fig. T10]. Create another cube and then set cube's **Size X, Size Y,** and **Size Z** fields to **25.591, 11.417,** and **1**, respectively.

3. Press **Ctrl+A** to select both the cubes. Choose **Arrange Objects | Center** from the **Tools** menu. On the **Attribute Manager | Options** tab, choose **Negative** from the **Z Axis** drop-down. Click **Apply** on the **Tool** tab to align objects [see Fig. T11].

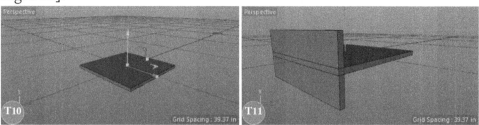

4. On the **Tool** parameter group, click **New Transform** and then choose **None, Positive, None** from the **X Axis, Y Axis,** and **Z Axis** drop-downs, respectively to align the two objects [see Fig. T12].

5. Select **Cube.1** from **Object Manager** or in the editor view and then align it with the bottom face of the cube using the **Move Tool** [see Fig. T13]. You can switch to the **Right** viewport to align the cubes accurately.

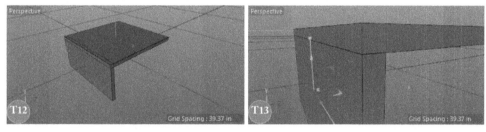

♡ *Tip: Aligning Objects*
*You can also quickly align objects using the **Axis Extension** feature of the Move tool, see video **m1vid-03.mp4** for reference.*

6. Make sure **Cube.1** is selected and then choose **Arrange Objects | Duplicate** from the **Tools** menu. In the **Attribute Manager | Duplicate** tab, enter **1** in the **Copies** field to create one duplicate of the cube.

7. Choose **Linear** from the **Mode** drop-down in the **Options** parameter group. On the **Position** section, turn off the X and Y switches. Enter **24.591** in the **Move [Z]** field to move the duplicate on the other side [see Fig. T14].

Now, we need to create the back support for the seat.

8. Create another cube in the editor view and then set its **Size X, Size Y,** and **Size Z** fields to **1, 25.591,** and **25.591,** respectively. Align the cube to the back of the seat using the process described above [see Fig. T15]. Similarly, create a cube for the font section and align it [see Fig. T16].

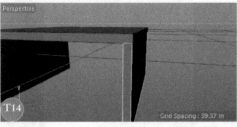
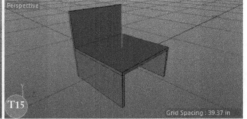

9. Press **Ctrl+A** to select all objects and then choose **Objects | Group Objects** from the **Attribute Editor's** menu. You can also press **Alt+G** to group the objects. Double-click on **Null** and rename it as **oneSeat**.

What just happened?
*Here, I've grouped objects [all cubes] in **Attribute Manager**. A **Null** object is created and selected objects are placed inside **Null**. Groups help you in better organizing your scene and keep **Object Manager** neat and tidy. When you group objects with children, chid objects are also placed inside **Null** and object hierarchies are maintained. You can expand a group using the **Shift+G** hotkeys. The objects that are one level below the parent are moved to the same level as the parent and the existing hierarchies are preserved.*

Creating Corner Section of the Sofa

Follow the steps given next:

1. Make sure **oneSeat** is selected and then create a copy of **oneSeat** in **Attribute Manager** or editor view [see Fig. T17].

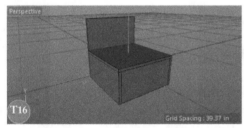
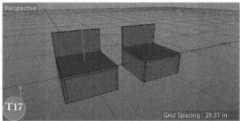

2. Rename the new group **Null** as **cornerSeat** in **Object Manager**. On **Object Manager**, select **cornerSeat | Cube.1_copies | Cube.1.0**. On the **Attribute Manager | Object** parameter group of **Cube.1.0**, set **Size X** and **Size Y** to **27.591** and **25.591**, respectively. Now, align the cube [see Fig. T18]. Similarly, create **cornerSeat** for the other end [see Fig. T19].

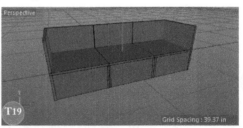

Tutorial 2: Creating a Coffee Table

In this tutorial, we will model a coffee table using the **Cylinder** and **Torus** primitives [see Fig. T1].

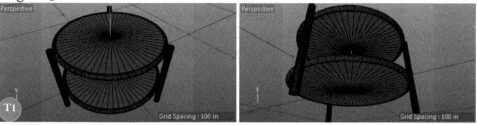

The following table summarizes the tutorial:

Table T2
Flow: The following sequence will be used to create the coffee table:

Flow: The following sequence will be used to create the coffee table:

1. Create the table top using the **Cylinder** and **Torus** primitives. **2.** Duplicate **Cylinder** and **Torus** to create bottom parts of the table. **3.** Create leg using the **Cylinder** primitive and then create copies using the **Array** function.

Skill level	Beginner
Time to complete	30 Minutes
Topics in the section	• Getting Started • Creating the Coffee Table
Project Folder	**chapter-m1**
Units	**Inches**
Final tutorial file	**m1-tut2-finish.c4d**

Getting Started

Start a new scene in CINEMA 4D and set units to **Inches**.

Creating the Coffee Table

Follow the steps given next:

1. Press and hold the LMB on **Cube** 🔲 on the **Standard** palette and then click **Cylinder** 🔲 . Press **NB** to enable **Gouraud Shading (Lines)** display mode. On the **Attribute Manager | Cylinder | Object** tab, set **Radius, Height,** and **Rotation Segments** to **37.5, 2,** and **60,** respectively.

2. Press and hold the LMB on **Cube** 🔲 on the **Standard** palette and then click **Torus** ⊙ . On the **Attribute Manager | Torus | Object** tab, set **Ring Radius, Ring Segments, Pipe Radius** to **37.5, 60,** and **1.521,** respectively. Align the two objects [see Fig. T2]. Select both objects and then press **Alt+G** to group the two objects and rename the group's **Null** to **tableTop**.

What next?
Now, we will create a clone of **tableTop** to create the bottom part of the table.

3. Invoke the M**ove** tool ✛ and ensure that **tableTop** is selected. Choose **Enable Quantizing** from the **Snap** menu or press **Shift+Q**. Drag the green Y-axis handle about **30** units in positive Y direction with **Ctrl** held down [see Fig. T3]. To enter accurate value, enter **30** in the **Position [Y]** field of **Coordinate Manager**. In **Object Manager**, rename the group as **tableBottom**. Press **Shift+Q** to disable the quantizing function.

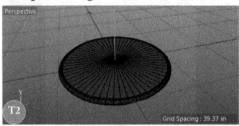

What just happened?
I have used the quantizing function. This function restricts stepless movement to a defined grid. For example, instead of a stepless rotation, you can allow rotation on **45** degrees increment. This function is primarily intended to use with the **Move** ✛, **Scale** ⬚, and **Rotate** ⟳ tools. However, you can use it with other tools such as **Polygon Pen** ✑.

You can specify settings for the quantizing function from the modeling settings. In **Attribute Manager**, choose **Modeling** from the **Mode** menu and then choose the **Quantize** tab. Select the **Enable Quantizing** check box and specify the quantizing settings using the **Movement**, **Scaling**, and **Texture** attributes.

4. Press and hold the LMB on **Cube** ⬛ on the **Standard** palette and then click **Cylinder** ⬜. Rename the cylinder as **leg**. On the **Attribute Manager | leg | Object** tab, set **Radius**, and **Height** to **2** and **50**, respectively. On **Coordinate Manager**, set **Position [Y]** and **Position [Z]** fields to **10**, and **40**, respectively to align leg with the **tableBottom** and **tableTop** [see Fig. T4].

What next?
Now, its time to create two more copies of the leg.

5. Ensure leg is selected in **Object Manager** and then **Alt** click on **Array** on the **Standard** palette to add an **Array** ❀ object. By default, the **Array** object creates **7** copies [see Fig. T5]. Also, notice that the **Array** object has used center of leg as a pivot to arrange the copies. Now, we will fix it.

What just happened?
*If you press **Alt** while selecting an object from the **Standard** palette, the selected object will become child of the new object. Otherwise, both object will be at the same level in **Attribute Manager**.*

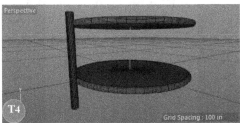

6. On the **Attribute Manager | Array | Coord** tab, set **P Z** to **0**. On the **Attribute Manager | Array | Object** tab, set **Copies** and **Radius** to **2**, and **40.5**, respectively.

Tutorial 3: Creating a Foot Stool

In this tutorial, we will model a foot stool using the **Cube** and **Cylinder** primitives [see Fig. T1].

The following table summarizes the tutorial:

Table T3	
Flow: The following sequence will be used to create the foot stool: **1.** Create two **Cube** primitives to create the seats of the foot stool. **2.** Use the **Cylinder** primitive to create legs.	
Skill level	Beginner
Time to complete	30 Minutes
Topics in the section	• Getting Started • Creating the Foot Stool
Project Folder	**chapter-m1**
Units	**Inches**
Final tutorial file	**m1-tut3-finish.c4d**

Getting Started

Start a new scene in CINEMA 4D and set units to **Inches**.

Getting the Foot Stool

Follow the steps given next:

1. Press **NB** to enable the **Gouraud Shading (Lines)** mode. Click **Cube** 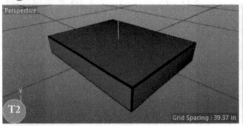 on the **Object** command group to add a cube to the editor view. Rename **Cube** as **baseGeo**. On the **Attribute Manager | baseGeo | Object** tab, set **Size X, Size Y,** and **Size Z** to **31.5, 5,** and **24.8,** respectively. Select the **Fillet** check box and set **Fillet Radius** and **Fillet Subdivision** to **0.2** and **3,** respectively [see Fig. T2].

2. Create another copy of the **baseGeo** by **Ctrl** dragging. Rename it as **topGeo** and place it on top of **baseGeo** [see Fig. T3]. Select **topGeo** and on the **Attribute Manager | topGeo | Object** tab, set **Size Y** to **8.** Align the two objects [see Fig. T4].

> **What next?**
> *Now, we will create legs for the foot stool.*

3. Click **Cylinder** on the **Object** command group to add a cylinder in the editor view. Rename it as **legGeo**. On the **Attribute Manager | legGeo | Object** tab, set **Radius** and **Height** to **1.5** and **4,** respectively. On the **Attribute Manager | legGeo | Caps** tab, turn select the **Fillet** check box and then set **Segments** and **Radius** to **3** and **0.5,** respectively. Align, **legGeo** with **baseGeo** [see Fig. T5]. Create three more copies of **legGeo** and align them.

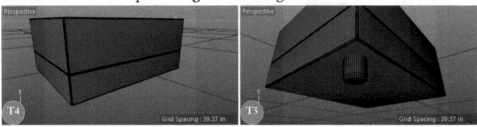

> **What next?**
> *Let's now arrange objects in layers.*

4. Click **Layer** switch on the right of the **topGeo** and choose **Add to New Layer** [see Fig. T6] from the popup menu; a new layer is created with the name **Layer** and **topGeo** is added to it. Also, CINEMA 4D assigns a color to the layer.

5. Press **Shift+F4** to open the **Layer Manager**. On **Layer Manager**, double-click on **Layer** and rename it as **mainGeoLayer**. Drag **baseGeo** from **Object Manager**

to the **mainGeoLayer** layer on **Layer Manager**. The **baseGeo** is now part of the layer **mainGeoLayer**.

6. Choose **File | New Layer** from the **Layer Manager's** menubar and rename the new layer as **legGeoLayer** [see Fig. T7]. Drag **legGeoLayer** from **Layer Manager** to the **legGeo** in the editor view to make **legGeo** part of the layer. Notice that the color of layers is now reflected in the editor view which helps in identifying which object is part which layer. Now, drag **legGeoLayer** on **legGeo.1**, **legGeo.2**, and **legGeo.3**. Double-click on the color swatches in the **Layer Manager** to change the color of the layers.

There are many toggle switches on right of the layer name. The following table summarizes these switches:

Table T3.1	
Icon	**Description**
S	Solo objects
V	Visible in editor
R	Visible in render
M	Show in managers
L	Lock layer
A	Animation on/off
G	Generators on/off
D	Deformers on/off
E	Switches XPresso, C.O.F.F.E.E. tags, etc. on or off.
X	Update/load XRefs

Tutorial 4: Creating a Bar Table

In this tutorial, we will model a bar table using the **Cube** and **Cylinder** primitives [See Fig. T1].

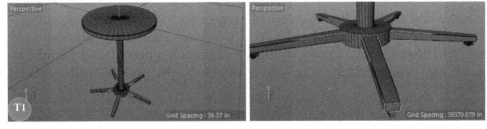

The following table summarizes the tutorial:

Table T4
Flow: The following sequence will be used to create the bar table: 1. Use a **Cylinder** primitive to create table top then use a **Cylinder** primitive and a **Pipe** primitive to create the vertical support for the table top. 2. Use the **Cube** primitives to create legs. 3. Use the **Taper** modifier to shape the leg. 4. Create wheels using a **Cube** primitive and a **Cylinder** primitive. 5. Apply the **Bend** deformer on the **Cube** object to give it shape of a wheel cover.

Skill level	Beginner
Time to complete	30 Minutes
Topics in the section	• Getting Started • Creating the Bar Table
Project Folder	**chapter-m1**
Units	**Inches**
Final tutorial file	**m1-tut4-finish.c4d**

Getting Started
Start a new scene in CINEMA 4D and set units to **Inches**.

Creating the Bar Table
Follow the steps given next:

1. Press **NB** to enable the **Gouraud Shading (Lines)** mode. Click **Cylinder** 🗌 on the **Object** command group to add a cylinder in the editor view. Rename it as **topGeo**. On the **Attribute Manager | topGeo | Object** tab, set **Radius, Height,** and **Rotation Segments** to **13.78, 1.5,** and **50,** respectively. On the **Attribute Manager | topGeo | Caps** tab, select the **Fillet** check box and then set **Segment** and **Radius** to **5** and **0.15,** respectively.

2. Create another cylinder for the central part and rename it as **centerGeo**. On the **Attribute Manager | centerGeo | Object** tab, set **Radius, Height,** and **Rotation Segments** to **1.3, 38,** and **18,** respectively. Align the two cylinders [see Fig. T2].

3. Click **Tube** on the **Object** command group to add a tube in the editor view. Rename the tube as **tubeGeo**. On the **Attribute Manager | tubeGeo | Object** tab, set **Inner Radius, Outer Radius, Rotation Segments,** and **Height** to **1.3, 4, 50,** and **2,** respectively. Align the **tubeGeo** at the bottom of the **centerGeo** [see Fig. T3].

What next?
Now, we are going to create support for the table.

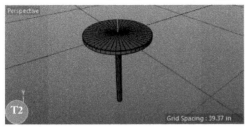

4. Create a cube in the viewport and then rename it as **supportGeo**. On the **Attribute Manager | supportGeo | Object** tab, set **Size X, Size Y, Size Z** to **1.6**, **12.8**, and **2.1**, respectively. Also, select the **Fillet** check box and then set **Fillet Radius** and **Fillet Subdivisions** to **0.1** and **3**, respectively.

5. Ensure **supportGeo** is selected in **Object Manager** and then on the **Standard** palette | **Deformer** command group, press and hold **LMB** on **Bend** . Click **Taper** with the **Shift** held down.

What just happened?
*Here, I have applied the **Taper** deformer to **supportGeo**. Holding down **Shift** ensures that the **Taper** object will be child of **supportGeo** otherwise it will be added at the same level as **suppportGeo**. The deformer objects modify the geometry of the other objects. You can use the deformer objects with the primitive objects, generators, polygon splines, and splines. Keep the following in mind while working with deformers:*

- *The deformer object only affects its parent.*
- *You can apply a number of deformer objects on an object.*
- *The order of the deformers is also important.*
- *The deformer objects are evaluated from top to bottom.*
- *The deformer objects have its origin and orientation.*
- *All deformers are activated automatically when you create them. A deformer object has no effect if it is deactivated.*

6. On the **Attribute Manager | Taper | Object** tab, click **Fit To Parent**. Set **Strength** to **40** and select the **Fillet** check box.

Tip: Taper Strength
You can interactively change the strength in the viewport by dragging the orange line [see Fig. T4].

7. Align the **supportGeo** with **tubeGeo** [see Fig. T5].

→ **What next?**
Now, we will work on the roller.

8. Create a cube and then set its **Size X, Size Y, Size Z** to **0.3, 0.926**, and **0.6**, respectively. Set **Segment Y** to **15**. Apply a **Bend** 🍰 deformer to the cube using the **Shift** key.

9. On the **Attribute Manager | Bend | Object** tab, click **Fit to Parent** and then set **Strength** to **180**. Now, align the cube [see Fig. T6]. Now, create a cylinder and align it [see Fig. T7]. Group the cube and cylinder that you just created with the name **rollerGrp**. Group **rollerGrp** and **supportGeo** with the name **baseGrp**.

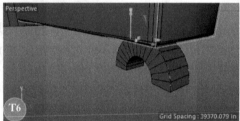

10. Switch to the **Top** view and ensure the **baseGrp** is selected. Press **L** to invoke the **Enable Axis** mode ⌊ and then move the axis at the center of the **topGeo** using **Move Tool** ✛ [see Fig. T8]. Press **L** again to disable the **Enable Axis** mode. Enable snapping for easily aligning axis.

? **What just happened?**
*Here, I've moved the origin of the group to the center of the model so that when I create copies of the **baseGrp**, they are rotated correctly around the new axis center. You can quickly enable/disable the **Enable Axis** ⌊ mode using the **L** key. Keep the following in mind:*

- *When you rotate or move axes of a hierarchical object, all axes of the child objects will also get affected.*
- *Before animating objects, ensure that you define the axis because if you rotate the parent, error will occur in the animation tracks of the child. The error occurs because of the change in the axes of the parent object.*
- *You cannot move axis while working with primitive objects. You need to make the object editable [first selecting it and then pressing **C**]. The workaround for primitive is that you make it child of a **Null** object and then move axis. The*

*quickest way to enclose an object inside a **Null** is that, select the object and then press **Alt+G**.*

- *Do not make multiple selections in **Object Manager** when you are temporarily making axis changes.*

11. Now, using the **Array** object 🎲 from the **Modeling** command group to create four more copies of the **baseGrp** [see Fig. T9].

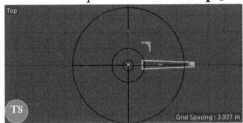

Quiz

Evaluate your skills to see how many questions you can answer correctly.

Multiple Choice
Answer the following questions, only one choice is correct.

1. Which of the following keys are used to invoke the **Live Selection** tool?

    [A] 8   [B] 9
    [C] 6   [D] 1

2. Which of the following mouse-drag operations is used to interactively change the size of the **Live Selection** tool's brush size?

    [A] LMB   [B] MMB
    [C] RMB   [D] Ctrl+RMB

3. Which of the following key combinations is used to render the current active view?

    [A] Ctrl+R   [B] Alt+Shift+R
    [C] Shift+R   [D] None of these

4. Which of the following key combinations is used to render the active view in the **Picture Viewer** window?

    [A] Ctrl+R   [B] Ctrl+Alt+R
    [C] Shift+R   [D] Alt+R

5. Which of the following key combinations is used to invoke the **Render Settings** window?

    [A] Ctrl+A   [B] Ctrl+B
    [C] Shift+A   [D] Alt+B

Fill in the Blanks

Fill in the blanks in each of the following statements:

1. The **X-Axis / Heading, Y-Axis / Pitch**, and **Z-Axis / Bank** tools are used to _____ or _____ the transformations along the X, Y, and Z axes, respectively.

2. The _____ key is used to make an object editable.

3. The _____ key combination is used to enable snapping.

4. The _____ key combination is used to display the project settings.

5. You can also maximize a viewport by _____ clicking on it.

6. The _____ hotkeys is used to invoke the **Commander** window.

7. The _____ hotkeys is used to invoke **Layer Manager**.

8. You can expand a group using the _____ hotkeys.

True or False

State whether each of the following is true or false:

1. When you resize a docked window, the surrounding windows are resized automatically.

2. You can select multiple views by clicking on the blank gray area of their headers with **Shift** held down.

3. The **Configure** option in the **MEV | Options** menu is used to specify the viewport settings.

4. The **Alt+G** key combination is used to group objects.

5. If you hold down the **Shift** key while dragging and dropping a layer onto an item in **Layer Manager**, the layer is assigned to the items children as well.

Practice Activities

Activity 1: Creating a Road Side Sign

Create a model of a road side sign [see Fig. A1].

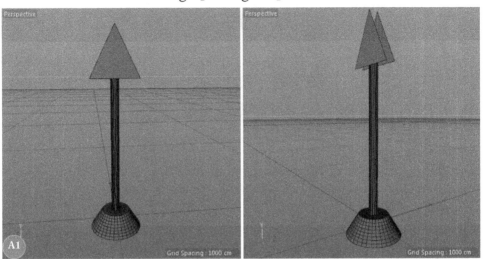

Hint Activity – 1
*Create the model using the **Cone**, **Cylinder**, and **Polygon** primitive objects.*

Activity 2: Creating a Robo

Create a robo model [see Fig. A2].

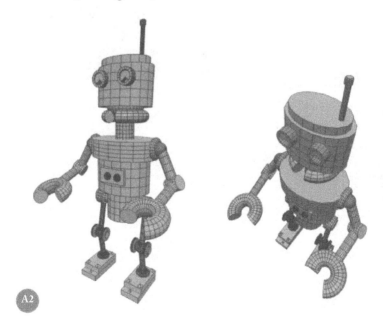

Hint Activity – 2
*Create the model using the **Cube**, **Sphere**, **Cylinder**, **Pyramid**, **Cone**, **Torus**, and **Tube** primitive objects.*

Activity 3: Coffee Table
Create the coffee table model [see Fig. A3] using the **Cube** primitive.

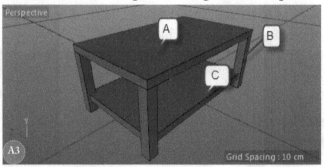

Activity 3 Dimensions:
A: *Length=35.433", Width=21.654", Height=1.5"*
B: *Length=34.037", Width=20.8", Height=1.5"*
C: *Length=2", Width=2", Height=13.78"*

Activity 4: 8-Drawer Dresser
Create the 8-Drawer Dresser model [see Fig. A4] using the **Cube** primitive.

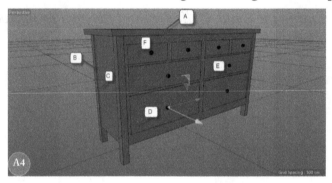

Activity 4 Dimensions:
A: *Length=65", Width=21", Height=1.5"*
B: *Length=2", Width=2", Height=35"*
C: *Length=60.76", Width=18.251", Height=30"*
D: *Length=27.225", Width=19.15", Height=11"*
E: *Length=27.225", Width=19.15", Height=7"*
F: *Length=12.871", Width=19.15", Height=5"*

Activity 5: Foot Stool

Create the foot stool model [see Fig. A5] using the **Cylinder** and **Box** primitives.

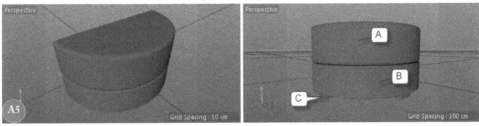

Activity 5 Dimensions:
A: Radius=14", Height=5.91", and Fillet=0.32"
B: Radius=14", Height=7.5", and Fillet=0.74"
C: Length=1.651", Width=3.455", Height=1.496", and Fillet=0.087"

Summary
In this chapter, the following topics are covered:

- Navigating the workspace
- Customizing the interface
- Understanding UI components
- Setting preferences
- Understanding layouts
- Moving, rotating, and scaling objects
- Managers, and Browsers
- Getting help

This page is intentionally left blank

- Creating guides in the editor view
- Interactively placing lights and adjusting their attributes in the scene
- Measuring angles and distances
- Working with Workplanes
- Arranging, duplicating, and randomizing objects
- Correcting lens distortions
- Creating virtual walkthroughs

Chapter M2: Tools of the Trade

In the last chapter, you learned how to create and place objects in the scene view. In this chapter, I will describe how you can place and arrange these objects accurately using arrange tools, guides, and **Workplanes**. You will learn to add annotations to the objects in the editor view so that you can mark objects in a complex scene for easy identification. Moreover, you will learn to create the virtual walkthrough. CINEMA 4D offers many tools that let you accomplish complex challenges with ease. These tools are available in the **Tools** menu of the main menubar. This chapter deals with these tools.

Guide Tool

 Guide Tool allows you to interactively create guidelines in the viewports. You can use handles of these guidelines to snap other entities such as vertices to them. You can create guidelines in one of the following ways:

Click on a viewport to create a handle, click again to create the second handle, and then press **Esc** to create a guide [see Fig. 1]; a red line appears. Now, you can click-drag the red line to create a guide surface perpendicular the view [see Fig. 2].

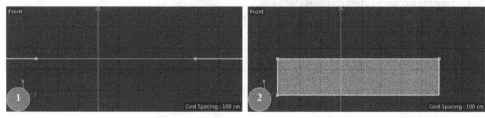

Click on a viewport to create a handle, click again to create the second handle. Don't release the mouse button and drag the mouse pointer and then click to create the guide surface.

To create a segmented guideline, click on a viewport to create a handle, click again to create the second handle. Now, a clicking within these handles creates a segmented guide [see Fig. 3]. To create a duplicate guide, click-drag one of the handles of the guide with **Ctrl** held down [see Fig. 4].

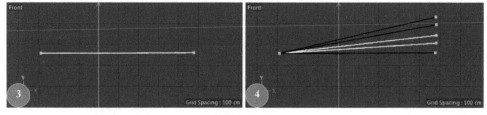

 Tip: The Delete Enabled Guides tool
The **Delete Enabled Guides** *tool deletes all guidelines which are active in* **Object Manager***. The active guidelines have green tick mark next to them [see Fig. 5]. No prior selection is required to delete the active guidelines.*

 Tip: Guides and Snaps
The guides work in combination with all **Snap** *options.*

Lighting Tool

This tool allows you to interactively create, select, and place a light object in the viewports. Also, it allows you to adjust the brightness of the lights without using **Attribute Manager**. The process is given next:

Invoke **Lighting Tool** from the **Tools** menu. Click on the empty area of the editor view to create a light source. Each click in the empty area of the editor will create a new light. Each new light will inherit the properties of the previously created light.

If lights already exist in the scene, most relevant light for a surface will automatically be marked when you place the mouse pointer on a surface. While manipulating the light, you can:

- Press **Shift** to move the light source in the direction of the current normals. You can control the distance between the light and surface using **Shift**.
- Press **Ctrl** to adjust the brightness of the light source.
- Press **Ctrl+Shift** to adjust the cone of the spot light.
- During manipulating light, press **Alt** to temporarily switch to the **Target** mode to adjust the target of the spot lights. For example, if you are using **Shift** to move the light along the direction of the current normals, you can press **Alt** while moving to temporarily switch to the **Target** mode and adjust the target of the light.

Caution: Lighting Tool
*The functionality of this tool is limited with the generators such as **Arrays** and **Cloners** unless you make the objects editable. When deformer objects are used, it might be cumbersome to locate the surface. In such cases, hide the **Deformer** object in the view.*

Naming Tool

T This tool allows you to efficiently rename object hierarchies in your scene. Although, this tool is specifically built for naming character rigs. However, you can use it to rename tag, material, layer, and take names. You can also use this tool to save already corrected named hierarchies as a preset and then use the preset to rename other hierarchies.

To understand working of this tool, create a series of five joints in the scene using **Joint Tool** and then rename them as **Hip, Knee, Foot, Ball**, and **Toe** [see Fig. 6]. Now, select **Hip** from **Object Manager** and then choose **Naming Tool** from the **Tools** menu. In **Attribute Manager**, click **Add** from the **Options** group. The **Name** dialog appears. In this dialog, type name as **leg** and click **OK**. Now, select **leg** from the **Type** drop-down. Enter **L_** in the **Prefix** field and **_$N** in the **Suffix** field. Now, click **Apply Name** to rename the objects [see Fig. 7]. The **$N** string is used to number objects automatically.

Now, for example, if you want to rename the hierarchy for the right leg, select all joints in **Object Manager**. In **Attribute Manager**, enter **L_** in the **Replace** field and **R_** in the **With** field. Click **Replace Name** to rename the objects [see Fig. 8].

Measure & Construction Tool

 You can use this tool to measure distance and angle between two objects. The measurement can be stored in a measurement object for later use. You can also adjust the distance and angles numerically after taking the measurement.

To measure distance, choose the **Measure & Construction** tool from the **Tools** menu and enable snapping. Press and hold **Ctrl+Shift** and drag the mouse pointer from one point to another point to measure distance [refer to Fig. 9].

To measure a new distance, click **New Measure** on the **Attribute Manager**. Repeat the process to measure the new distance. To measure an angle, create a second measurement line by **Ctrl** clicking on a point [see Fig. 10]. Now, you can change distance and angle directly in the viewport by dragging the value labels or from **Attribute Manager** using the **Distance 1**, **Distance 2**, and **Angle** attributes.

To move arrowheads, drag them with **Shift** [red arrowheads] or **Ctrl** [green arrowheads] held down. You can also use this tool to move individual points, edges, and polygons of an object.

Annotation Tool

 This tool is used to interactively create **Annotation** tags in the viewports [see Fig. 11]. These tag are very useful in adding object-specific comments to the objects in a complex scene. The text fields move with the corresponding object. If the object is not visible, the corresponding text fields will also not be visible in the viewport.

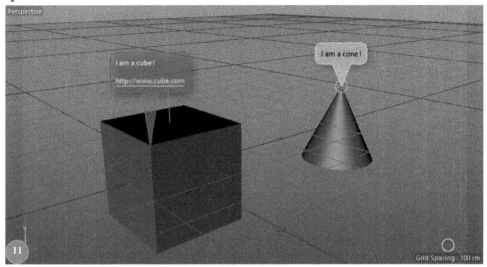

To create a tag, press and hold the LMB on the object, a red cursor appears. This cursor snaps to the valid object points. If you release the mouse button without snapping the cursor, the annotation is snapped to the center point of the object. You can press **Esc** to abort the creation process. When you create an annotation, an **Annotation** tag is applied to the object. Once you create an annotation, you can change the text and URL from the **Attribute Manager | Annotation** tab.

You can double-click on a field, to unfold as well as activate it. If you hover the mouse pointer on the annotations, the folded and scaled down boxes will be maximized temporarily.

Note: The WWW tag
*The older **WWW** tag will be automatically converted to the **Annotation** tags when loaded in CINEMA 4D.*

Workplanes

Workplanes are primarily designed for use in the technical modeling in which elements are arranged along perpendicular axis. They are generally used in the **Perspective** view. You can use **Workplanes** to place and arrange objects across a plane. However, the position, scale, and rotation will always be according to the world coordinate system. This system always remains visible as a light gray grid in a viewport.

Align Workplane to X, Align Workplane to Y, Align Workplane to Z

 You can use these commands to arrange the workplane to the respective axis. The normals will be always oriented along the Y-axis. The **Align Workplane to Y** mode is the most commonly used command, as it resembles the world grid.

Align Workplane to Selection

This command rotates the **Workplane** according to the currently selected element.

Align Selection to Workplane

This command positions the selected elements on the **Workplane**.

Arranging Objects

Cinema 4D offers many tools to arrange objects in the scene. Let's explore them.

Arrange

This command allows you to arrange, scale, or rotate selected objects. When you change the attributes in **Attribute Manager**, real-time feedback is displayed in the viewports. To arrange objects, select objects and then choose **Arrange Objects | Arrange** from the **Tools** menu. In **Attribute Manager**, set the desired **Mode**, and then click **Apply** if the objects are not instantly arranged in the viewport. Now, adjust the settings in **Attribute Manager** as long as the **Arrange** function is active, see Figs. 12 and 13.

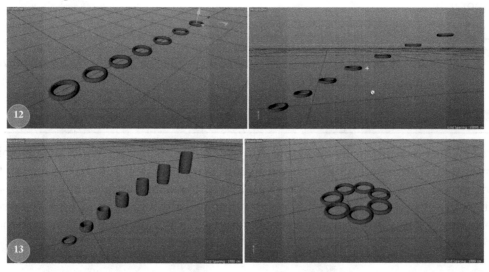

Center

This command allows you to center objects in the 3D space in the viewports. This command is applied to all selected objects in the **Object Manager**. However, the children of the selected objects are not affected. An object in CINEMA 4D is enclosed inside a cuboid of bounding box. The axis system used for alignment considers the center of the axis at the center of the bounding box. Refer to Tutorial 1 of Chapter M1 to understand functioning of this tool.

Duplicate

This command allows you to create as many as duplicates you want to create depending on the RAM available. You can also transform duplicates using this command. Most of the options available for this tool are similar to that of the **Arrange** command. Refer to Tutorial 1 of Chapter M1 to understand functioning of this tool.

Transfer

This command allows you to copy the **PSR** values [position, rotation, and scale] from one object to another. To transfer values, select the object[s] you want to modify and then execute the **Transfer** command. Now, hover the mouse over the source object, a white line appears. Click to transfer values.

Randomize

You can use this function to randomly place objects in 3D space. You can also randomize the scale and rotation of the objects. To randomize objects, select them and then execute the **Randomize** command [see Fig. 14].

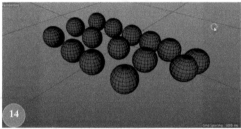

Lens Distortion

The **Lens Distortion** tool allows you to deal with the lens distortion effects in CINEMA 4D. You can correct the distortion in the image shot with the small focal lengths [shot with wide angles]. This distortion can be problematic while motion tracking a footage. The motion tracking produces better results when used with the distortion free footage. The distortions can be classified as barrel distortion, pincushion distortion, and mustache distortions. Fig. 15 shows the barrel, pincushion, and mustache distortion, respectively. The live footage generally have the barrel shaped distortions. You can use the **Lens Distortion** tool to create a lens profile that you can use later when tracking a live footage.

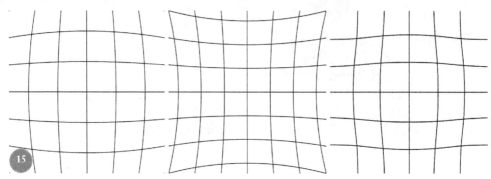

There are two algorithms that you can use to solve the distortion: **Manual** and **Automatic**. The **Manual** method is for the experienced users in which the tool's settings are manually adjusted. The **Automatic** method is the recommended method and works well in most of the situations.

You can use the **Lens Distortion** tool to create guides in the image. Try to match the guides with the curved lines [the lines that should be straight] on the image. You can **Ctrl+Drag** a guide to create duplicates. **Ctrl+Click** once on a guide to create a new point on the line. You can delete the selected point by using **Backspace** or **Del**.

Keep the following in mind while creating lens profiles:

- Place guides precisely. The more guides [precise] you use, the better result you get.
- The lens distortion is calculated accurately where the density of guide is more. However, the rears with low number of guides can be calculated incorrectly. Therefore, you need to place and spread out guides strategically.
- To restore the size of the original image after correction, you can modify the scale of the image as well as its offset values.

Doodle

The options available for **Doodle** allow you to sketch in the editor view. You can use this tool to mark corrections, make notes, load bitmaps, and so on. These options work well with a tablet, however, you can use a mouse as well. When you draw in the editor view, each drawing is saved in the doodle object for each frame.

 Tip: Rendering Doodle
*By default, the **Render Doodle** option is active in **Render Settings | Options**. As a result, the doodles appear in the render output and this is the reason you can only draw doodle in the **Render Safe** area of the viewport.*

Virtual Walkthrough

There are two tools available that allow you to walk or fly through the CINEMA 4D scenes. This experience is similar to a third-person shooter game where you can walk or fly though the scene. The camera path can be recorded or can be output as spline. A virtual walkthrough can be created in CINEMA 4D using one of the following methods:

- Use the **Virtual Walkthrough** tool to fly freely through the scene. To exit this mode, select any other tool.
- Use the **Collision Orbit** tool to define an orbital path around an object.
- Use the **Collision Detection** tag to exclude specific objects such as doors from the collision detection.

Keep in mind the following while working with these tools:

- If you stuck somewhere in the scene [due to collision detection] and not able to move, press **Ctrl** to temporarily deactivate the detection.
- These tools works with the polygonal objects. If you face any issue with the geometry, make it editable by first selecting it and then pressing **C**.
- These tools behave correctly if a floor [polygonal object] is used in the scene.

Collision Orbit Tool

 You can use this tool in conjunction with **Target Camera** to orbit around an object. The camera will automatically avoid the obstacles. To understand working of this tool:

- Create a **Target Camera** in the scene.
- Target the object [around which you want to orbit] using **Camera's Null**.
- Switch to the target camera.
- Invoke the **Collision Orbit** tool and move the mouse in the 3D view with LMB held down. The **HP** rotation values will be displayed in **HUD** in the viewport. The motion path can be recorded.

The collision is detected when the camera's z-axis [path of the view] intersects with the given object. A green icon appears when this happens.

Virtual Walkthrough Tool

 You can use this tool to fly around the scene. The mouse and keyboard can be used to navigate the scene as you do in a 3D third person shooter game.

3D Connexion

The option available for **3D Connexion** work with the 3D mouse sometimes also referred to as **Spacemouse**. This mouse offers six directions for movement. CINEMA 4D uses the 3DxWare driver to operate the **Spacemouse**. It is must that you install the driver before starting the 3D mouse.

Quiz

Evaluate your skills to see how many questions you can answer correctly.

Multiple Choice
Answer the following questions, only one choice is correct.

1. Which of the following tools is used to create guides interactively in the editor view?

    [A] Guide Tool   [B] Create Guide Tool
    [C] A and B   [D] None of these

2. Which of the following key combinations along with LMB is used to create distance arrow while measuring distances using the **Measure & Construction** tool?

    [A] Ctrl+Shift   [B] Ctrl
    [C] Alt   [D] Ctrl+Alt

Fill in the Blanks
Fill in the blanks in each of the following statements:

1. _____ are primarily designed for use in the technical modeling in which elements are arranged along perpendicular axis.

2. The _____ command allows you to center objects in the 3D space in the viewports.

3. The _____ command allows you to copy the PSR values [position, rotation, and scale] from one object to another.

4. You can use the _____ tool to create a lens profile that you can use later when tracking a live footage.

5. The _____ feature allow you to sketch in the editor view.

True or False

State whether each of the following is true or false:

1. You can create a segmented guide by using the **Alt** key.

2. **Guide Tool** works in combination with all **Snap** options.

3. **Lighting Tool** allows you to interactively create, select, and place a light object in the viewports.

4. **Naming Tool** allows you to efficiently rename object hierarchies in the scene.

5. The measurement computed using the **Measure & Construction** tool can not be stored in a measurement object for later use.

Practical Activity

Activity 1: Arranging Objects

Create shape of the word "Love" using splines and then use the **Cube** primitive, and the **Arrange** and **Duplicate** functions to create text, as shown in Fig. A1.

Hint Activity - 1

*Create a spline using the **Sketch** tool along which the cubes will be duplicated. Create another spline using the **Pen** tool that will be used to rotate the cubes, see Fig. A2. Use the **Duplicate** function to create **100** copies of the cube and then use the **Arrange** function to arrange duplicate cubes on the spline.*

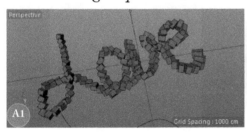
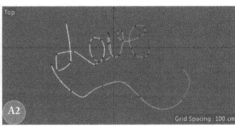

Summary

This chapter covered the following topics:

- Creating guides in the editor view
- Interactively placing lights and adjusting their attributes in the scene
- Measuring angles and distances
- Working with Workplanes
- Arranging, duplicating, and randomizing objects
- Correcting lens distortions
- Creating virtual walkthroughs

Chapter M3: Spline Modeling

A spline is a sequences of vertices, connected by lines [segments] lying in 3D space. However, a spline has no 3-dimnesional depth. Splines are infinitely thin entities and are not visible during rendering. However, if you are using Studio version of CINEMA 4D, you can render splines using Hair, and Sketch & Toon features. The shape of these lines is defined by the interpolation method. The interpolation method defines whether the shape of the line is curved or straight. The curved splines have a soft leading edge without any sharp corner. In this chapter, I will cover the spline modeling tools and techniques.

A spline is made up of several partial curves called lines or segments. You can create holes using splines, when a spline lies completely inside another spline and both splines are closed. If two segments overlap, no 3D surface will be created or you will see strange results.

CINEMA 4D offers a number of predefined parametric spline curves known as spline primitives. These primitives are calculated using the mathematical formulae and therefore they have no points [vertices] to edit. However, you can convert a parametric spline primitive to an editable spline by first selecting it and then pressing **C**.

To adjust vertices using the **Pen** tool [I will discuss is shortly], you must make the spline editable. However, you can apply **Deformers** to a spline even if the spline has not been made editable. You can access spline primitives from the **Spline** option of the **Create** menu or from the **Spline** command group from the **Standard** palette [see Fig. 1].

Working with Spline Tools

CINEMA 4D offers four tools for drawing splines: **Pen**, **Sketch**, **Spline Smooth**, and **Spline Arch Tool**.

Sketch Tool

You can use this tool to directly draw in the viewport and to edit the exiting splines. For example, you can use it to redirect, smooth, and connect splines. This tool obeys snap settings. Snap settings are particularly useful when you want to draw a curve on the surface of an object. This tool also works well with tablets.

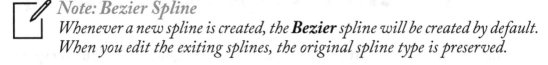

Note: Bezier Spline
*Whenever a new spline is created, the **Bezier** spline will be created by default. When you edit the exiting splines, the original spline type is preserved.*

To draw a spline using this tool, invoke it, and then click and drag in the editor view. As soon as you release LMB, spline will be created.
Some of the functions of this tool are given below:

- You can grab a spline from any location and edit it from that location.
- You can connect two splines [connected at their ends] or two segments of the same spline. If you connect the start and end points of a spline, the spline will

be closed. You can also connect two splines or two segments of the same spline anywhere along the spline.

- You can temporarily switch to the **Spline Smooth** tool by pressing the **Shift** key. The previously defined settings of the **Spline Smooth** tool will be used.
- If you want to unify spline segments or two separate splines, select the splines first. MMB drag to specify the radius of the brush interactively, and then drag in the view to connect the splines. Two separate splines will be combined into one.

Spline Smooth Tool

This tool not only allows you to smooth the splines but also offers a number of other features [works as a shaping brush] that can be combined seamlessly. You can adjust the strength and radius of this tool interactively by dragging the mouse pointer vertically or horizontally. This tool can create lots of points on the spline when you are using feature such as random. You can fix this by switching to the **Smooth** mode.

Note: Spline types
This tool works with all types of splines.

Spline Arch Tool

 You can use this tool to create variety of arc shapes and soft connections between segments. This tool creates 3 point circle and obeys snap settings. To create an arch three points are required: start point, middle point, and end point.

Pen Tool

 This tool is used to create or edit splines. It replaces the previous **Bézier**, **B-Spline**, **Linear** and **Akima Spline** tools and provides a wide variety of new functions. To create a spline, invoke the tool and then click on the view to create a point. If you drag the mouse pointer, a tangent is created to produce a curved section. A ghost preview of the next segment is shown before you create the next point. It helps in visualizing the next segment that will be created if you click to create next point.

Some of the functions of this tool are given below:

- This tool obeys snapping settings.
- The points created using this tool can be selected and moved without need of selecting them.
- If you click on a segment using this tool, the neighboring points will be selected.
- You can make multiple selections using **Shift**.
- When you click on a point or segment, the tangents are displayed and can be edited.
- If you want to extend a spline at the beginning or at the end, select the respective point and then click to create a new point. If a point is selected along the spline [other than the end point], press **Ctrl** to continue at the spline's end.
- If you position the mouse pointer on a segment, it will be highlighted. RMB

click shows a menu with number of options that you can use to edit splines.

- Splines points and segments can be moved by clicking and dragging on them. If you press **Shift** and then drag the tangent handles, the tangents will be broken and the selection will be manipulated between the neighboring points.
- Double-clicking on a point toggles between the soft tangent [both halves have same length] and null tangent.
- If you hover mouse over the spline with **Ctrl** held down, a preview appears that shows where a matching tangent will be inserted.
- If you connect the start and end points of a spline, the spline will be closed.
- You can also connect sections of a single spline. If the spline contains other open sections, a new spline will be created.

As mentioned above, if you RMB click on a point or segment, a popup appears with several options. The following table summarizes these options.

Table 1: The context popup options	
Option	**Description**
Delete Point	Deletes the selected point. You can also delete a point by **Ctrl** clicking on it.
Disconnect Point	Disconnects the spline at the selected point. If spline is a closed spline, the spline will be separated at the selected location and both end points will have the same shared location.
Hard Tangents	Creates an angular point by setting tangent length of the **Bezier** splines to **0**.
Soft Tangents	Sets equal length for the both halves of the tangent and creates a soft curve.
Kill Edge	Deletes the spline section.
Add Point	Adds a new point at the clicked location.
Split Point	Adds two new, non-coherent, congruent points at the clicked location of the segment. The spline will be split into sections.
Hard Edge	Tangents' length will be set to zero thus the spline will be made linear. Double-clicking on a point do the same.
Soft Edge	The spline will be made curve. Double-clicking on a point do the same.

Working with Generators

The tools available in the **Generators** command group are some of the most powerful tools of CINEMA 4D that allow you to create surfaces quickly, see Fig. 2. The **Generators** are interactive

and they use other objects to generate their surfaces. You can also access these tools from the **Generators** sub-menu of the **Create** menu. The following table summarizes the generators.

Table 2: The generators		
Option	**Icon**	**Description**
Subdivision Surface		This tool is one of the most powerful sculpting tools offered by CIENMA 4D to a digital artist. You can create almost any shape using the point weighting, and edge weighting. This tool uses an algorithm to subdivide and round the object. You can use any kind of object with this tool, however, most of the times, you will work with polygons.
Extrude		You can use this tool to extrude a spline to create an object with depth.
Lathe		When you apply this tool on a spline, it rotates the spline about the Y axis of the local axis system of the generator object to generate a surface of revolution.
Loft		Use this tool to stretch a skin over two or more splines.
Sweep		This tool works with two or three splines. The first spline referred to as contour spline, defines the cross section and is swept along the second spline, referred to as path. The optional third spline, referred to as rail spline, controls the scale of the contour spline over the object's length.
Bezier		This tool stretches a surface over Bezier curves in the X and Y directions.

Options in the Mesh Menu

Some of the spline tools, commands, and options are available in the **Mesh** menu. These can be accessed from the **Mesh | Spline** menu and are discussed next:

Hard Interpolation

This command changes all selected points to hard interpolation. If no points are selected, all points on the spline are changed to hard interpolation, see Fig. 3. When you execute this command, CINEMA 4D essentially sets the length of the tangents to **0**.

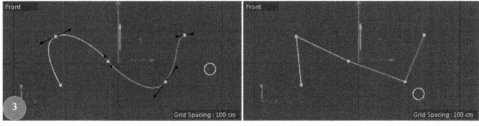

Soft Interpolation

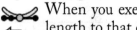 This command changes all selected points to soft interpolation, see Fig. 4.

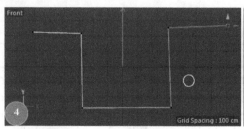 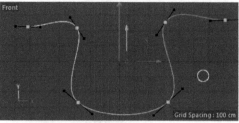

Equal Tangent Length

 When you execute this command, the shorter tangent handle is set to the length to that of the second associated tangent, see Fig. 5.

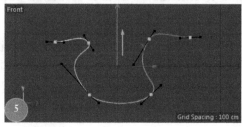 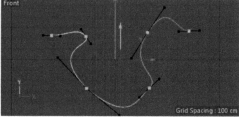

Equal Tangent Direction

This command is used to restore the smoothness of the broken tangents, see Fig. 6.

Tip: Breaking Tangents
*You can use the **Move** tool with the **Shift** key to break the tangent's handles association and move one tangent handle while leaving the other unchanged.*

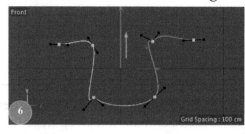 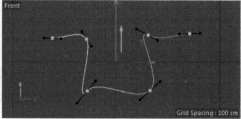

Join Segment

 You can use this command to connect several unconnected segments. To connect segments, select one or more points of each segment and then use the **Join Segment** command, see Fig. 7.

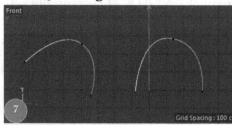

Break Segment

This command is used to create a new spline segment. To break spline, select one of the points and execute this command, a new segment appears and all points on either side of the separated segment will become a new segment, see Fig. 8. Note that this command works in the **Points** mode only.

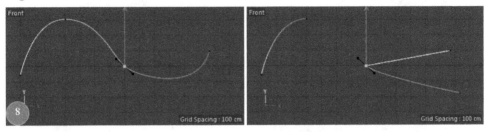

Explode Segments

You can use this command to split a spline into separate objects. For this command to work, you don't need to have a selection or the **Points** mode active. New spline objects become child of the original spline.

Set First Point

You can use this command to make the selected point first point of a spline. Note that the start of a spline is displayed in white color whereas the end displayed in the blue color. This command needs a point selection and it works in the **Points** mode only.

Reverse Sequence

Use this command to reverse the point order of a segment. This command can be applied to several segments of a spline by **Shift** selecting various segments.

Move Down Sequence

When you use this command, the sequence number of each point is incremented and the last point of the original sequence becomes the first point in the new sequence.

Move Up Sequence

When you use this command, the sequence number of each point is decremented and the last point of the original sequence becomes the first point in the new sequence.

Chamfer

This tool is an interactive tool that you can use to convert a selected point to two points with soft-interpolation between them, see Fig. 9. If no points are selected, the entire points on the spline will be chamfered.

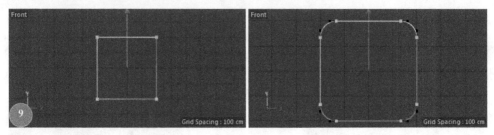

Create Outline

This tool is also an interactive tool and is used to create outline around the original spline, see Fig. 10.

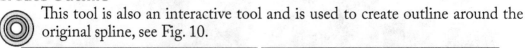

Cross Section

This tool is used to create a cross section for a group of spline, see Fig. 11. Before you use this tool, the splines must be selected in **Object Manager**.

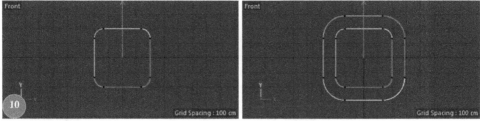

Line Up

Use this command to align sequentially selected points to a straight line, see Fig. 12.

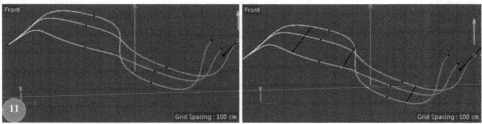

Project

You can use this command to project a spline onto surfaces of the other objects, see Fig. 13. To understand the working of this command, create a sphere and helix. Make helix spline primitive editable. Execute the **Project** command. From the **Options** parameter group in **Attribute Manager**, select **Spherical** from the **Mode** drop-down and then from the **Tool** parameter group, click **Apply**.

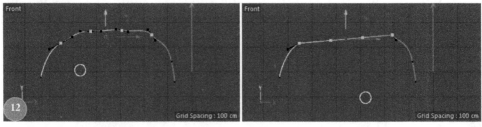

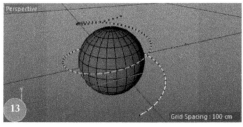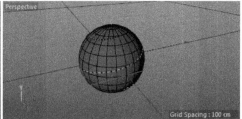

Round

You can use this command to round and subdivide the sequentially selected points of a spline, see Fig. 14.

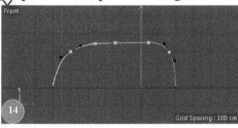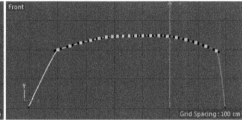

Spline Subtract

This command is used for subtract boolean operations. The surfaces that overlaps the target spline will be cut out, see Fig. 15.

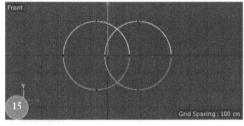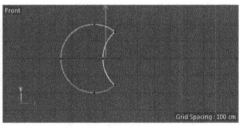

Spline Union

When you use this command, the splines are unified and overlapping surfaces are removed, see Fig. 16.

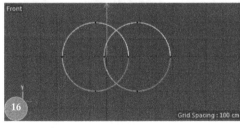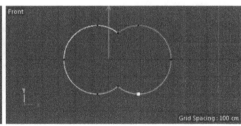

Spline And

When you use this command, a new spline is created out of the overlapping regions of all splines that are included in the operation, see Fig. 17.

Spline Or

The opposite of the **Spline And** function, see Fig. 18.

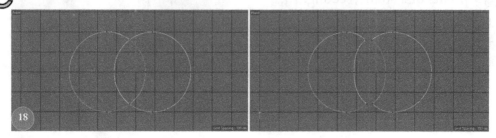

Spline Intersect

This command is a combination of the results of the **Spline And** and **Spline Or** commands, see Fig. 19.

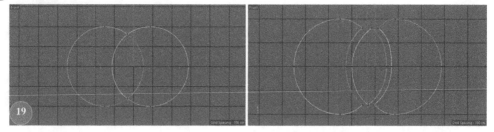

Note: Spline Tools and Commands
*See Video: **m3vid–01.mp4** to get better understand spline tools and commands.*

Tutorials

Before you start the tutorials, create a folder with the name **chapter-m3**. We'll use this folder to host all tutorial files and other resources.

Tutorial 1: Creating Model of a Pear

In this tutorial, we will model a pear using splines and **Lathe** generator [Fig. T1].

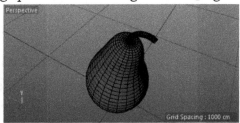

The following table summarizes the tutorial:

Table T1	
Flow: The following sequence will be used to create the pear model: **1.** Import **pear.jpg** to viewport background. **2.** Use the **Pen** tool to draw the profile of the pear. **3.** Use the **Lathe** generator to create surface of the pear. **4.** Use the **Cylinder** primitive and the **Bend** deformer to create stem.	
Difficulty level	Beginner
Estimated time to complete	20 Minutes
Topics	• Getting Started • Creating the Pear Model
Resources folder	
Tutorial units	**Centimeters**
Final tutorial file	**m3-tut1-finish.c4d**

Getting Started

Ensure that you have access to **pear.jpg** in the **chapter-m3** folder. Start a new scene in CINEMA 4D and set units to **Centimeters**.

Creating the Pear Model

Follow the steps given next:

1. Open **Windows Explorer** and drag **pear.jpg** to the **Front** view. Press **Shift+V** to open the **Viewport [Front]** options in **Attribute Manager**. In the **Attribute**

Manager | Back tab, set **Transparency** to **65%** and **Offset Y** to **435** so that bottom of the pear sits on the origin [Fig. T2].

2. Choose **Pen** ✐ from the **Standard** palette | **Spline** command group and then click and drag to the origin to create the first point. Follow the pear shape to create a profile curve [Fig. T3] and then press **Esc** to complete the creation of the curve. Don't worry about the exact placement of the points. You can always come back and adjust the points using the **Pen** tool.

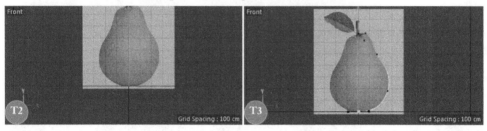

3. Select the bottom point and then enter **0** in the **Position [X]** and **Position [Y]** fields in **Coordinate Manager** to place the point at the origin point. Similarly, select the top point and set its **Position [X]** to **0**.

> ⓘ *What just happened?*
> *By positioning the points, we have ensured that the both points are in a line in 3D space. It will ensure that there will be no hole in the geometry when we will apply the **Lathe** generator in the next step.*

4. Ensure spline is selected and then click **Standard** palette | **Generators** command group | **Lathe** 🏺 with **Alt** held down to connect **Lathe** object to the **Spline**. Switch to the **Perspective** view to see the shape of the pear. Press **NB** to enable the **Gouraud Shading (Lines)** mode [Fig. T4].

5. Select **Spline** in **Object Manager** and then in the **Attribute Manager | Object** tab, choose **Natural** from the **Intermediate Points** drop-down. Now, set **Numbers** to **10** [Fig. T5].

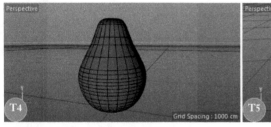 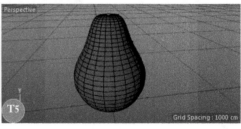

> ✎ *What just happened?*
> *The options in the **Intermediate Points** drop-down define how the spline is further subdivided with the intermediate points. This setting only affects when spline is used with the generators. When you select the **Natural** method for interpolation, the **Number** field corresponds to the number of intermediate*

points between vertices. The points on the curvature are positioned closer together where is the spline has more curvature.

6. Ensure **Lathe** is selected in **Object Manager** and then click **Bend** 🍥 from the **Standard** palette | **Deformer** command group, to add a **Bend** object. Ensure the order of the objects, as shown in Fig. T6. On the **Attribute Manager | Bend | Object** tab, click **Fit to Parent**. Set **Size [Y]** to **430** and **Strength** to **-16** [Fig. T7].

What next?
Now, we will create a stem for the pear.

7. Create a **Cylinder** 🛢 primitive. On the **Attribute Manager | Cylinder | Object** tab, set **Radius, Height, Height Segments**, and **Rotation Segments** to **6, 79, 24,** and **24,** respectively. Align it to the top of the cylinder [Fig. T8]. Apply a **Bend** deformer to it and then on the **Attribute Manager | Bend | Object** tab, click **Fit to Parent**. Set **Strength** to **-120** to bend the stem [Fig. T9].

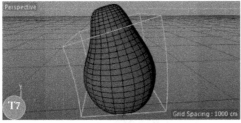

Note: Subdivisions
*If you want to increase the number of subdivisions along the rotation of the curve, select the **Lathe** object and then on the **Attribute Manager | Lathe | Object** tab, change the value of the **Subdivision** field. You can use the **Isoparm Subdivision** to define the number of isoparms used to display the **Lathe** object when the isoparms display mode is active. Press **NI** to activate this mode. Fig. T10 shows the pear mode with **Isoparm Subdivision** set to **12**.*

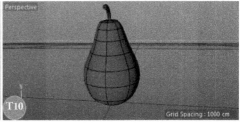

Tutorial 2: Creating a Glass Bottle and Liquid

In this tutorial, we will model a glass bottle and liquid using the **Lathe** generator [Fig. T1].

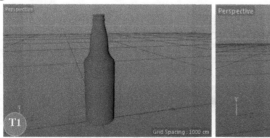 

The following table summarizes the tutorial:

Table T2	
Flow: The following sequence will be used to create the glass bottle and liquid: 1. Use a background image to draw profile of the bottle using the **Pen** tool **2.** Create surface using the **Lathe** generator. **3.** Use the **Split** function to create liquid.	
Difficulty level	Intermediate
Estimated time to complete	35 Minutes
Topics	• Getting Started • Creating the Bottle • Creating the Liquid
Resources folder	**chapter-m3**
Tutorial units	**Centimeters**
Final tutorial file	**m3-tut2-finish.c4d**

Getting Started

Ensure that you have access to **beer.jpg** in the **chapter-m3** folder. Start a new scene in CINEMA 4D and set units to **Centimeters**.

Creating the Bottle

Follow the steps given next:

1. Open **Windows Explorer** and drag **beer.jpeg** to the **Front** view. Press **Shift+V** to open the **Viewport [Front]** options in **Attribute Manager**. In the **Attribute Manager | Back** tab, set **Size X** and **Size Y** to **400**. Also, set **Transparency** to **65%** and **Offset Y** to **188** so that bottom of the bottle sits on the origin. Create a shape using the **Pen** tool [Fig. T2]. Rename the spline as **bottle**. Select the bottom point and set its X and Y positions to **0**. Select the top point and set its X position to **0**.

2. RMB click on bottle in editor view. Choose **Create Outline** from the menu and then drag the spline to create an outline [Fig. T3]. Notice that the **Create Outline** function created a closed spline. On the **Attribute Manager | bottle | Object** tab, clear the **Close Spline** check box. Select the second point from the bottom and make sure its **X** position is set to **0** [Fig. T4].

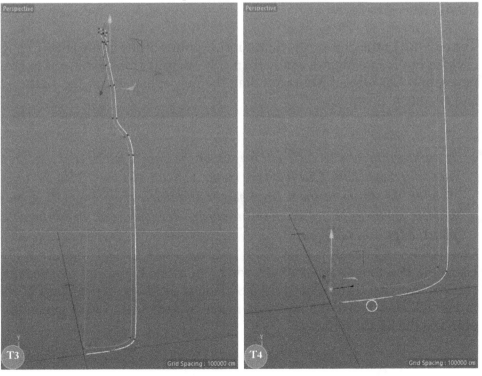

3. Select the points at the top [Fig. T5] and then press **Delete** to make a straight shape inside of the bottle. Now, apply a **Lathe** generator  to create shape of the bottle [Fig. T6].

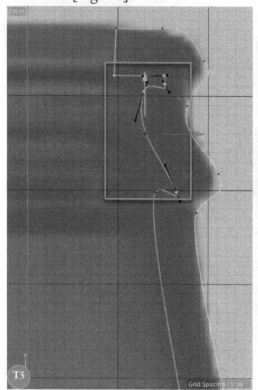

Creating the Liquid

Follow the steps given next:

1. Turn off the **Lathe** object . Select inside points of the bottle [Fig. T7]. RMB click on bottle in editor view and then choose **Split** from the menu to create a new spline. On **Object Manager**, drag the new spline out of the **Lathe** group and rename it as **liquid**.

> *Tip: Liquid Level*
> *If you want to create a different height for the liquid, RMB click on the spline and then choose **Line Cut** from the menu. You can also press **MK**. Now, create a cut on the spline on the desired height and delete the points. The **Line Cut** tool lets you to cut polygon and spline objects. The tool works in all three modes: **point**, **edge** and **polygon**.*

2. Using the **Pen** tool , click on the top point of the liquid and then extend the curve to the vertical green line [Fig. T8]. Adjust the shape of the lqiuid, as shown in Fig. T9. Now, create the liquid geometry by applying a **Lathe** generator object to the **liquid** spline.

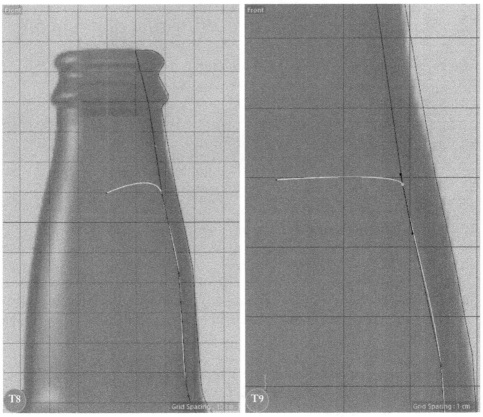

Tutorial 3: Creating a Martini Glass

In this tutorial, we will model a martini glass using **Lathe** generator [Fig. T1].

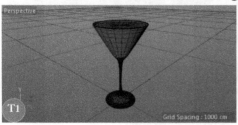
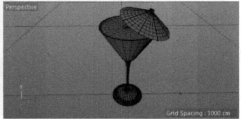

The following table summarizes the tutorial:

Table T3	
Flow: The following sequence will be used to create the martini glass: **1.** Create the glass and liquid, as done in tutorials 2. **2.** Create the umbrella object using the **Cone** primitive. **3.** Use the **Box** primitives and the **Array** function to create sticks.	
Difficulty level	Intermediate
Estimated time to complete	45 Minutes
Topics	• Getting Started • Creating the Glass and Liquid • Creating the Umbrella
Resources folder	**chapter-m3**
Tutorial units	**Centimeters**
Final tutorial file	**m3-tut3-finish.c4d**

Getting Started

Ensure that you have access to **martini-glass.jpg** in the **chapter-m3** folder. Start a new scene in CINEMA 4D and set units to **Centimeters**.

Creating the Glass and Liquid

Follow the steps given next:

1. Open **Windows Explorer** and drag **martini-glass.jpg** to the **Front** view. Press **Shift+V** to open the **Viewport [Front]** options in the **Attribute Manager**. In the **Attribute Manager | Back** tab, set **Size X** and **Size Y** to **300** and **456.472**, respectively.

2. Set **Transparency** to **65%** and **Offset Y** to **194** so that bottom of the bottle sits on the origin. Also, set **Offset X** to **2**. Create a shape using the **Pen** tool [Fig. T2]. Now, create the geometry for the glass and liquid, as described in tutorial 2 [Fig. T3].

Creating the Umbrella

Follow the steps given next:

1. Create a cone object ⌂ in the scene. On the **Attribute Manager | Cone | Object** tab, set **Top Radius, Bottom Radius**, and **Height** to **10, 100**, and **57**, respectively. On the **Cap** tab, clear the **Caps** check box. Create a cube and then on the **Attribute Manager | Cube | Object** tab, set **Size X, Size Y**, and **Size Z** to **120, 8**, and **8**, respectively. Align it with the cone [Fig. T4].

3. Select cube in **Object Manager** and press **Alt+G** to group it inside a **Null** ⌐⁰ object. Press **L** and then move the axis to the center of the cone [Fig. T5]. Press **L** again to disable **Enable Axis** ⌐ mode.

4. Ensure **Null** is selected in **Object Manager** and then click **Array** 🔅 on the **Standard** palette with **Alt** held down. On the **Attribute Manager | Array | Object** tab, set **Radius** to **0** and **Copies** to **8**. Select **Cube** on **Object Manager** and then press the **R** key to activate the **Rotate** tool. Now, rotate the objects along Z-axis [Fig. T6]. You can now modify the dimensions of the cube and cone so that they better fit with each other.

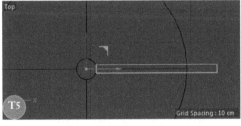

5. Create a **Cube** primitive and a **Cylinder** primitive to create the stick and then align the stick with the umbrella.

Tutorial 4: Creating the Apple Logo

In this tutorial, we will create apple logo in 3D, as shown in Fig. T1.

Table T4
Flow: The following sequence will be used to create the apple logo:
1. Import image to viewport background and create the profile curve using the **Pen** tool. **2.** Use the **Mirror** function to create other half of the profile. **3.** Create **Circle** spline to draw the cut. **4.** Use the **Spline Subtract** and **Spline And** functions to create final shape. **5.** Use the **Extrude** generator to create surface.

Difficulty level	Beginner
Estimated time to complete	40 Minutes
Topics	• Getting Started • Creating the Logo
Resources folder	**chapter-m3**
Tutorial units	**Centimeters**
Final tutorial file	**m3-tut4-finish.c4d**

Getting Started

Follow the steps given next:

1. Start a new scene in CINEMA 4D and set units to **Centimeters**. Maximize the **Front** viewport and then press **Shift+V** to open the viewport settings.

2. On the **Attribute Manager | Back** tab, assign **apple-logo.jpg** to the **Image** parameter. Also, set **Offset X, Offset Y**, and **Transparency** to **8, 398**, and **90**, respectively.

Creating the Logo

Follow the steps given next:

1. Create a shape using the **Pen** tool in the **Front** view [Fig. T2]. Make sure that the X coordinate value of the two end points is **0**. Press **Ctrl+A** to select all points and then choose **Transform Tools | Mirror** from the **Mesh** menu.

2. Click drag towards the left in **Front** view; a line appears in the view. Snap the line to the two end vertices [Fig. T3] and then release the mouse button to create the mirror copy [Fig. T4].

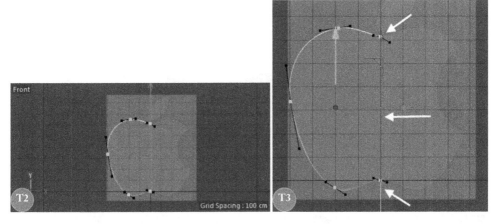

3. Select the top points [Fig. T5], RMB click and then choose **Soft Interpolation** from the popup menu. Weld the two points using the **Weld** command. Make sure **Spline** is selected in the **Object Manager** and then in the **Attribute Manager | Object** tab, make sure the **Close Spline** check box is selected.

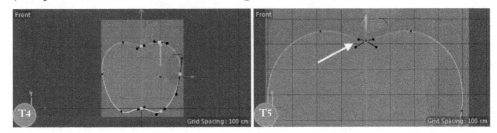

> ✎ *Note: Gap in spline*
> *If you see a wierd looking spline around the bottom points [see Fig. T6], use the **Pen** tool to select them. RMB click on them and then choose **Delete Point** from the shortcut menu.*

4. Create a **Circle** object, align it with the **Spline** [Fig. T7] and then press **C** to make it editable. On **Object Manager**, select **Circle** and then select **Spline** [select them in the correct order]. Choose **Spline | Spline Subtract** from the **Mesh** menu to create the shape shown in Fig. T8.

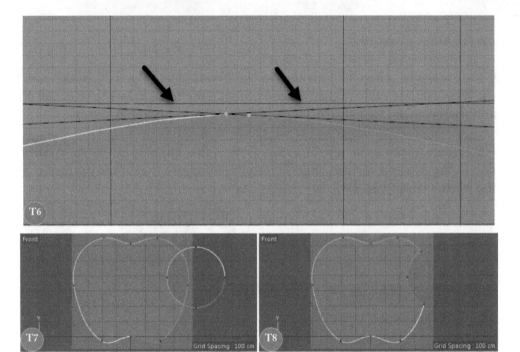

5. Create two circles, make them editable and then align, as shown in Fig. T9. Select the two circles and then choose **Spline | Spline And** from the **Mesh** menu to create the shape shown in Fig. T10.

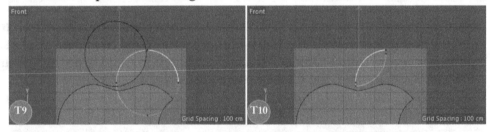

6. Select both splines in **Object Manager** and then choose **Conversion | Connect Objects + Delete** from the **Mesh** menu to combine them. Now, connect the unified spline with the **Extrude** generator.

7. On the **Attribute Manager | Object** tab, set **Movement** Z to **30**. On the **Caps** tab, set **Start** and **End** to **Fillet** Cap. Also, set **Step** and **Radius** for **Start** and **End** to **3** and **1.5**, respectively.

Tutorial 5: Creating a Whiskey Bottle

In this tutorial, we will create a whiskey bottle using the **Loft** generator [Fig. T1].

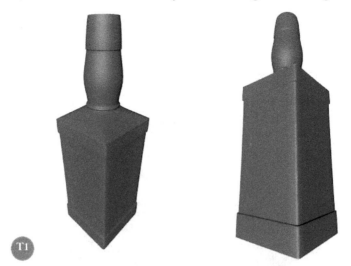

The following table summarizes the tutorial:

Table T5	
Flow: The following sequence will be used to create the whiskey bottle: **1.** Use the **Rectangle** and **Circle** splines to follow the bottle's shape. **2.** Use the **Loft** generator to create surface.	
Difficulty level	Beginner
Estimated time to complete	40 Minutes
Topics	• Getting Started • Creating the Bottle
Resources folder	**chapter-m3**
Tutorial units	**Centimeters**
Final tutorial file	**m3-tut5-finish.c4d**

Getting Started

Follow the steps given next:

1. Start a new scene in CINEMA 4D and set units to **Centimeters**. Maximize the **Front** viewport and then press **Shift+V** to open the viewport settings.

2. On the **Attribute Manager | Back** tab, assign **whiskey.jpg** to the **Image** parameter. Also, set **Offset X**, **Offset Y**, and **Transparency** to **0**, **381**, and **90**, respectively.

Creating the Bottle

Follow the steps given next:

1. Choose **Rectangle** from the **Standard** palette | **Spline** tab to create a rectangle in the editor view. In the **Attribute Manager | Rectangle | Object** tab, set **Width** and **Height** to **232** and **232**, respectively.

2. Turn on the **Rounding** switch and then set **Radius** to **8.248** and **Plane** to **XZ**. Align the rectangle in the **Front** view [Fig. T2]. Create a copy of the rectangle and align it [Fig. T3].

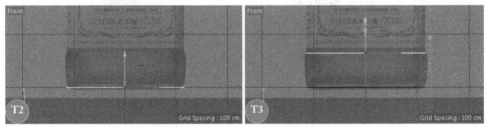

3. Similarly, create five more rectangles and align them [Fig. T4]. You need to adjust the size of the rectangles to get the shape you are looking for.

4. Choose **Circle** from the **Standard Palette | Spline** tab to create a circle in the editor view. In the **Attribute Manager | Circle | Object** tab, set **Radius** to **54**, **Plane** to **XZ**, and then align it [Fig. T5]. Similarly, create more circles, adjust their radii and then align them [Fig. T6]. Fig. T7 shows all the rectangles and circles in the scene.

5. Now, connect all shapes with the **Loft** generator to create shape of the bottle. Ensure the order of the shapes is as shown in Fig. T8.

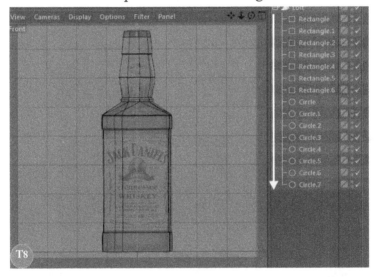

Quiz

Multiple Choice

Answer the following questions, only one choice is correct.

1. Which of the following tools uses an algorithm to subdivide and round the objects?

    [A] Extrude   [B] Lathe
    [C] Loft   [D] Subdivision Surface

2. Which of the following tools is used to stretch a skin over two or more splines?

[A] Extrude   [B] Lathe
[C] Loft   [D] Subdivision Surface

3. Which of the following key combinations is used to activate the isoparm display mode?

[A] NI   [B] NP
[C] NB   [D] ND

4. Which of the following key combinations is used to activate the **Line Cut** tool?

[A] MI   [B] MP
[C] MK   [D] MD

5. Which of the following keys is used to activate the **Enable Axis** mode?

[A] **E**   [B] **A**
[C] **C**   [D] L

Fill in the Blanks
Fill in the blanks in each of the following statements:

1. You can convert a parametric spline primitive to an editable spline by first selecting it and then pressing _____.

2. _____ creates 3 point circle and obeys snap settings.

3. The options in the _____ drop-down define how the spline is further subdivided with the intermediate points.

True or False
State whether each of the following is true or false:

1. Whenever a new spline is created, the **Bezier** spline will be created by default.

2. The **Extrude** tool is used to extrude a spline to create an object with depth.

Practice Activities

Activity 1: Creating a Bowl
Create a model of bowl using the **Lathe** generator [see Fig. A1]. Use **bowl.jpg** as reference.

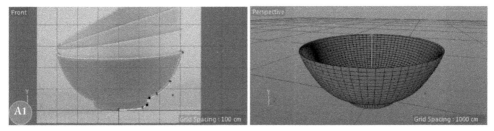

Activity 2: Creating a Candle Stand
Create the candle stand model [see Fig. A2] using the **Lathe** generator.

Activity 3: Creating a Glass Table
Create model of a glass table using the **Rectangle** spline, **Cylinder** primitive, and **Extrude** generator [see Fig. A3].

Activity 4: Creating a Corkscrew
Create model of a corkscrew using the **Helix** spline and **Sweep** generator object [see Fig. A4].

Summary

This chapter covered the following topics:

- Splines tools
- Splines functions
- Generators command group

Chapter M4: Polygon Modeling

Polygons are a type of geometry that you can use to create 3D models in CINEMA 4D. Polygons represent and approximate surfaces of a 3D model. Many 3D modelers use the primitive objects [discussed in Chapter M1] as the basic starting point for creating models and then create complex geometries using the sub-object levels [components] such as points, edges, and polygons. You can also apply commands such as **Bevel**, **Extrude**, **Bridge**, and so on, on a primitive's polygon mesh in order to modify the primitive's shape.

Closed shapes in a plane with three or more sides are called polygons. The endpoints of the sides of polygons are called points or vertices. The line connecting two points is called an edge. Polygons are classified by how many sides or angles they have. Following list shows types of polygons based on number of sides they have:

- A triangle is a three sided polygon.
- A quadrilateral is a four sided polygon.
- A pentagon is a five sided polygon.
- A hexagon is a six sided polygon.
- A septagon or heptagon is a seven sided polygon.
- An octagon is an eight sided polygon.
- A nonagon is a nine sided polygon.
- A decagon is a ten sided polygon.

There are a variety of techniques that you can use to create 3D polygonal models in CINEMA 4D but before we start creating models, let's first understand the tools, commands, modes, and options that are required to build a 3D model. Let's first start with different modes used in polygonal modeling in CINEMA 4D.

Working with Modes

CINEMA 4D provides three modes for polygonal modeling. These modes are: **Points**, **Edges**, and **Polygons** and you can access them from the **Tools** palette. Before you use these modes, you need to make a primitive object editable. To do so, ensure that the object is selected and then press **C**. When you make an object editable, you loose its parametric creation attributes.

Points

Use the **Points** mode when you want to edit points of an object. When this mode is active, the points appear as small squares. Hover the cursor on the points to highlight them. To select points, you can use the selection tools. You can also select them by clicking on points one by one. The selected points appear in yellow color.

To add points to the selection, click on them with **Shift** held down. To remove points from selection, **Ctrl** click on them. To select all points, choose **Select All** from the **Edit** menu or press **Ctrl+A**. To de-select all points, choose **Deselect All** from the **Edit** menu or press **Ctrl+Shift+A**. To delete selected points, choose **Delete** from the **Edit** menu. Alternatively, you can press the **Backspace** or **Delete** key.

Edges

Use this mode to edit the edges of the polygons, selected edges are highlighted in color. You can select edges much the same way as you select points.

Polygons

In CINEMA 4D, you can work on three types of polygons: triangles, quadrangles, and n-gons. You can select polygons much the same way as you select points or edges.

 Note: Using Transformation tools
*You can also use the **Move**, **Scale** and **Rotate** tools to edit the selected edges, points, or polygons. To add to the selection, hold **Shift** while you make selection. To remove from selection, click on the object with **Ctrl** held down.*

Selecting Objects and Components

In order to make models in any 3D application, you should be able to select objects or sub-objects/components [points, edges, and polygons]. CINEMA 4D offers various tools and commands for making selections. Let's explore these tools and commands available in the **Select** menu. These options are also available in the **U** hidden menu.

Selection Filter

The options in the **Selection Filter** sub-menu allow you to choose which types of object can be selected in the viewport. By default, all objects in the viewport are selectable. These options are very useful in scenes where you have many lights, cameras, bones, and so on. For example, if you are working on the lighting of the scene, you can make only light selectable and disable all other by first choosing **Selection Filter | None** from the **Select** menu and then choosing **Selection Filer | Light** from the **Select** menu.

Loop Selection

Hotkey: UL

Loops are elements that are connected in a shape of a loop. The **Loop Selection** tool allows you to quickly select the loops. Fig. 1 shows the loop selection in **Polygons**, **Edges**, and **Points** modes.

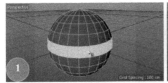  

Tip: Loop Selection
*If you are in the **Edges** mode, you can quickly select a loop by double-clicking on an edge using the **Move**, **Select**, or **Rotate** tool. If a point or polygon is already selected in one of the modes, press **Ctrl+Shift** and then click on the next element to create a selection between it and the already selected element.*

Tip: Handling
*You can influence of the length of the loop by dragging instead of clicking when the **Loop Selection** tool is active. If you just want to select the boundary loops, select the **Select Boundary Loop** check box from **Attribute Manager**. If you select the **Stop at Boundary Edges** check box, the loop will stop at boundary edges [it works in the **Points** and **Edges** modes].*

Ring Selection
Hotkey: UB

The function of this tool is similar to that of the **Loop Selection** tool, however, it select the elements that form a broad ring-shape.

Outline Selection
Hotkey: UQ

In the **Polygons** mode, it selects the edges that outline the selected polygons. To select outline edges, move the mouse pointer over the polygon selection. When the edges that outline the selection change color, click to select edges, see Fig. 2. Also, the **Edges** mode gets activated. This tool also works in the **Edges** mode.

 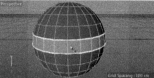 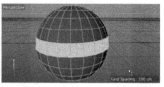

Fill Selection
Hotkey: UF

In the **Edges** mode, this tool creates a polygon selection from an existing edge selection, see Fig. 3. To fill the selection, move the mouse pointer over the edge selection. When the polygons change color, click to select those polygons. Also, the **Polygons** mode will get activated.

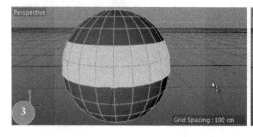 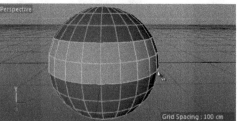

Path Selection
Hotkey: UM

This tool lets you select polygon edges or points by painting on the edges or points. This tool works only in the **Edges** or **Points** mode, see Fig. 4. Click and drag over elements to create a selection.

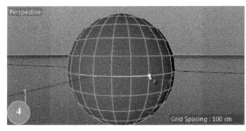 

Phong Break Selection
Hotkey: UN

 This tool works with the low-res mechanical models whose edges are assigned via **Phong Break Shading**.

 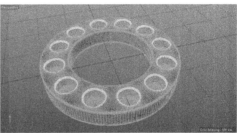

Select All

Use this command to select all points, edges, or polygons of the currently selected object.

Deselect All

Use this command to de-select all points, edges, or polygons of the currently selected object.

Invert
Hotkey: UI

Use this command to invert the current selection.

Select Connected
Hotkey: UW

Use this command to select all points, edges, or polygons connected to the selected element, see Fig. 6.

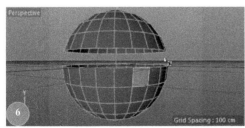 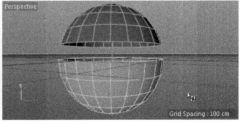

Grow Selection

Hotkey: UY

 You can use this command to add to the selection. All adjacent elements (depending on the mode selected) are added to the selection, see Fig. 7.

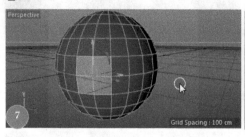 

Shrink Selection

Hotkey: UK

Use this command to remove from the selection, see Fig. 8.

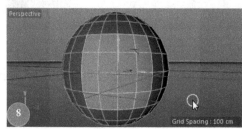 

Hide Selected

Hides the currently selected elements.

Hide Unselected

Hides all unselected elements.

Unhide All

Makes all hidden elements visible again.

Invert Visibility

 Use this command to make visible elements hidden and hidden elements visible.

Convert Selection

Hotkey: UX

With this command, you can convert one type of selection to another. On executing this command, the **Convert Selection** dialog appears. Select the desired options from the dialog and then click **Convert** to convert the selection, see Fig. 9.

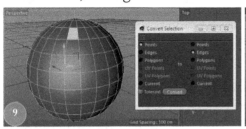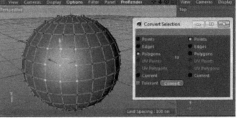

Set Vertex Weight

Use this tool to set the vertex weight. To create a vertex map, select the points or polygons and then execute this command, the **Set Vertex Weight** dialog appears. In this dialog, set the value for the **Value** attribute and then click **OK**, see Fig. 10.

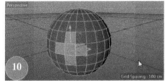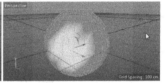

Set Selection

You can use the **Set Selection** command to store selection sets and then recall them later. You can create selection sets for point, edge, and polygon selections. Make a selection and then execute this command, a tag is added to **Object Manager**. You can manipulate the selection from **Attribute Manager**, see Fig. 11.

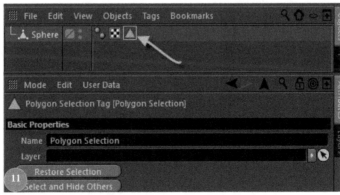

Note: Selection Filter

*The commands in the **Select | Selection Filter** submenu are used to specify which types of the object will be displayed in the viewport. By default, all objects are selectable in the viewport. Note that you can still select objects in **Object Manager**.*

The **Select | Selection Filter | Selector** *command allows you to select all lights, nulls, texture tags, and so on. For example, if you want to select all lights in the viewport and then want to make their* **Intensity** *value to* **0**. *Choose* **Select | Selection Filter | Selector** *to open the* **Selector** *dialog. Now, select* **Light** *from the dialog; all light objects will be selected. Now, in* **Attribute Manager***, set* **Intensity** *to* **0***. Similarly, you can select tags using the* **Tags** *tab of the* **Selector** *dialog.*

Adjusting Structure of the Polygonal Objects

The tools available in the **Mesh** menu are used to change the structures of the polygonal objects. Most of these tools are available in the **Points, Edges,** and **Polygons** modes. These tools work on the editable objects. You can make an object editable by choosing **Conversion | Make Editable** from the **Mesh** menu or by pressing **C**. Generally, these tools affect the selected points, edges, or polygons. However, if no component is selected, these tools affect the entire selected object.

✏️ *Note: Primitive Objects*
The primitive objects and splines objects in CINEMA 4D are parametric; they have no points or polygons. They are created using parameters and math formulae. To edit these objects at component level, you need to first convert them to points or polygons. You can do so by pressing **C** *from the keyboard, clicking the* **Make Editable** *icon on the* **Tools** *palette, or by choosing* **Conversion | Make Editable** *from the* **Mesh** *menu.*

✏️ *Note: Inactive tools*
Any structure tool that cannot be used on the current selection will be grayed out. For example, if you make a point selection, the **Mesh | Create Tools | Edge Cut** *tool will be grayed out. If you are using an interactive tool, the most recently action can be undone by pressing* **Esc** *as long as the mouse button is still pressed.*

💡 *Tip: Using hotkeys*
When you are modeling, you can temporarily activate a function using hotkeys. For example, if you select polygons using the **Live Selection** *tool, press and hold* **D** *to temporarily activate the* **Extrude** *tool. Extrude the polygons and then release the hotkey to switch back to the* **Live Selection** *tool.*

💡 *Tip: Modeling popup menu*
You can also quickly access the structure tools from the RMB popup menu. The options available in the menu depend on the type of component selected. You can also quickly access these tools from the **U** *and* **M** *hidden menus.*

Several tools have their own specific parameters that you can access from the **Options** tab of the tool's **Attribute Manager**, see Fig. 12. The options in the **Attribute Manager | Tool** tab allow you to choose whether the changes will be applied automatically in real-time or not. If you are working on a heavy scene and your system is responding slowly to the automatic updates, you can disable this feature by clearing the **Realtime Update** check box from the **Attribute Manager | Tool** tab, see Fig. 13.

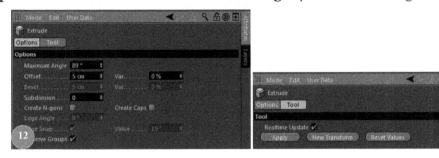

If you clear the **Realtime Update** check box, you need to click on the **Apply** button to apply the changes to the object. If you want to reapply the tool, click **New Transform**. You can also use this button to repeatedly apply changes to the object. For example, if you are using the **Extrude** tool, clicking repeatedly on this button will extrude the selected element(s) multiple times. The **Reset Values** button can be used to reset the tool to its default state. The tools, commands, and options available in the **Mesh** menu are discussed next.

Conversion Sub-menu

The following commands are available in the **Conversion** sub-menu:

Make Editable

Hotkey: C

The primitives objects in CINEMA 4D are parametric and created using math formulae. These primitive objects have no points, edges, or polygons. To create complex objects using these primitives, you need to first make the primitive objects editable. When you make an object editable, you get access to the object's components: points, edges, and polygons. You can use the **Make Editable** command to make an object editable. Note that this command is one-way. You cannot convert an editable object back to a parametric object.

Current State to Object

This command allows you to collapse the stack for the selected object and creates a polygon copy of the selected object. For example, if you have applied multiple deformers on an object, you can use this command to create a polygon copy of the resulting shape. If you apply this command to a parametric object, it will create a polygon copy considering all deformers.

Caution: Child Objects
This command ignores child objects, therefore, you need to apply this command separately for each child object.

Caution: Animation Data
The animation data is not copied to the new object when you use this command.

Connect Objects

You can use this command to create a single object from multiple objects. When you connect the polygonal objects to which you have applied materials and selection tags, CINEMA 4D ensures that selection tags are connected properly. Also, texture projections are restored accurately. The original objects are preserved, you can delete them if they are no longer required in the scene.

Caution: Animation Data
The animation data is not copied to the new object when you use this command. The original objects and their animation data is preserved.

If you are connecting splines of different types, the new resulting spline will be a **Bezier** spline. When you use this command, not only the connected object occupies less space in **Object Manager**, it also renders quickly even though it has same number of polygons.

Connect Objects+Delete

The function of this command is similar to that of the **Connect Objects** command but additionally, it deletes the original objects.

Polygon Groups to Objects

You can use this command to create separate polygonal objects from the from the non-connected surfaces [polygons groups], see Fig. 14.

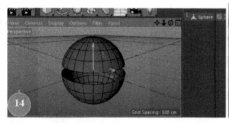 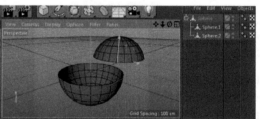

Commands Sub-menu

The following commands are available in the **Commands** sub-menu:

Array

This command allows you to duplicate selected elements [points or polygons] of an object and then distribute them randomly in the 3D world, see Fig. 15. You can also vary size as well as rotation of the duplicates. If no elements are selected, all points and surfaces of the selected object are duplicated. You can use this command, for example, to create a complete meadow from a single blade of polygon grass.

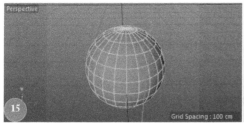
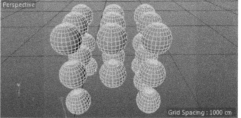

Clone

Use this command to create duplicate of the surface or points of an object. You can then rotate the duplicated objects along the object axis. The duplicates shown in Fig. 15 are created from a cube. The settings use to create the duplicate are shown in the right image of Fig. 16.

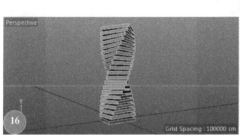
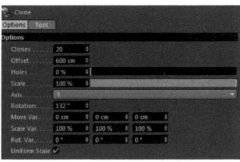

Disconnect

Hotkeys: UD, Shift+D

This command is used to disconnect the selected polygons from the object or segments between the selected spline points, see Fig. 17. The disconnected surfaces will remain in place but no longer will be connected to the original element. This command can be applied on splines. Unlike the **Break Segment** tool, the start and end points of the disconnected segments are duplicated and are not deleted from the original spline. The order of the spline remains intact both before and after the disconnection. If you press **U** followed by **Shift+D**, the **Disconnect** dialog appears. In this dialog, you can find the **Preserve Groups** setting. If you clear the check box, the elements are separated from each other as well as from the original object. If this check box is selected, the elements are disconnected from the original object in one piece.

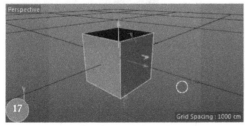
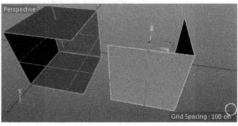

Split
Hotkey: UP

This command is little bit different than the **Disconnect** command. When you apply this command, the disconnected surfaces create a separate object leaving the original object unchanged. This tool can also be applied on splines. For splines, a point selection or **Points** mode is required.

Collapse
Hotkey: UC

This command collapses the selected points, edges, or polygons to a single center point. These points can be welded together, see Fig. 18.

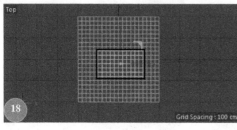
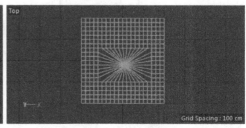

Connect Points/Edges
Hotkey: MM

This command works in the **Points** and **Edges** modes and connects points and edges, see Figs. 19 and 20.

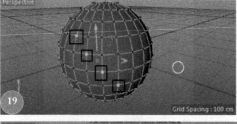
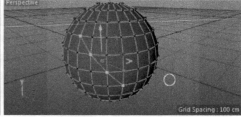
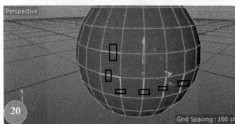
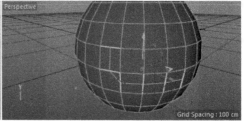

Melt
Hotkey: UZ

As name suggests, this function melts the selected points, edges, or polygons. Fig. 21 shows the melted point, edges, and polygons, respectively.

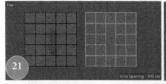

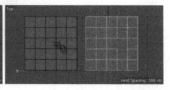

Dissolve

 Use this command to delete selected edges. This command works similar to the **Melt** command. However, it deletes the unnecessary points as well. It is ideal for deleting the unnecessary edges created by the **Connect Points/ Edges** command. When you delete the points the Phong angle is also taken into account. However, if you execute this command with **Shift** held down, it deletes all unnecessary points regardless of the Phong angle. In the **Points** and **Polygons** modes, this command exactly works like the **Melt** command.

Subdivide
Hotkeys: U~S, Shift+S

This command is used to subdivide polygon objects or splines, see Fig. 22. If no elements are selected, it subdivides the whole object. To define the subdivision level, click on the gear icon next to this command [or press **U**, followed by **Shift+S**], the **Subdivide** dialog appears. In this dialog, specify a value for the **Subdivision** attribute and then click **OK**.

Triangulate

This command converts the polygons into triangles, see Fig. 23. You might need this command if you are exporting mesh to an application that takes only triangulated geometry.

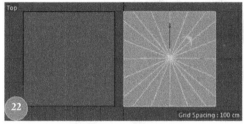
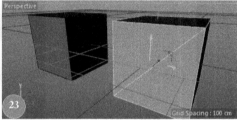

Note: Triangles
Generally, try to use quads as much as possible during the modeling process. Quads take less memory, they render faster, and produce better shading when used with the subdivision surface.

Untriangulate
Hotkeys: UU, Shift+U

If you are importing a geometry built only of triangles, you can use this command to convert the triangles into quadrangles. Triangles that cannot be converted into quads are left in their original state. You can specify the settings for this command from the **Untriangulate** dialog that opens when you click the gear icon located next to this command. The **Evaluate Angle** check box in the dialog lets you specify the angle at which the resulting quad will be created between two triangles.

Set Point Value
Hotkey: MU

This command allows you to set values for the selected points. You can use this command to center, quantize, or crumple points. This command works on the polygon points, spline points, and FFD points.

Spin Edge
Hotkey: MV

This command is used to spin the selected edges and then connect them to the neighboring two points, see Fig. 24.

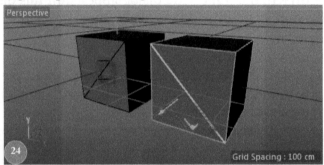

Edge to Spline

Use this command to create a spline from an edge selection, see Fig. 25.

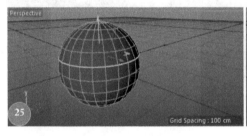
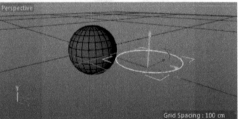

Change Point Order

A polygon is either a triangle or a quadrangle. A triangle has points A, B, and C whereas a quadrangle has four points A, B, C, and D. If these four points are not on the same plane, the quadrangle is a non-planar polygon. The **Change Point Order** command lets you change the order of the points of a polygon. For example, the **Matrix Extrude** tool orients itself to the coordinate system which refers to the order of the points A, B, C, and D.

Optimize
Hotkey: UO, Shift+O

When you create objects using the **Connect Objects** command, very often, some points and surface would be duplicated in the resulting geometry. You can use this command to remove these points. You can also apply this tool to spline points.

Reset Scale

You can use this command to restore the coordinate system of the object. You need to do this when the object's axes are not perpendicular to each other or have different lengths. Note that this type of issue occurs in CINEMA 4D's versions prior to R12.

Modeling Settings
Hotkey: Shift+M

When you select this option, the **Modeling Settings** section appears in **Attribute Manager**. From this section, you can specify various modeling settings such as **Snap** and **Quantize**.

Create Tools Sub-menu
The following commands are available in the **Create Tools** sub-menu:

Create Point
Hotkey: MA

This tool allows you to add new points to the objects. There is no need to make a selection and this tool works in all three modes. To add a point on an element, move the mouse pointer over the element, the element appears highlighted. Click on element to add the point. If you don't release the mouse button, you can drag the point to change its position on the element. This tool also works with splines.

Polygon Pen
Hotkey: ME

The **Polygon Pen** tool is a super tool, it is more than just a polygon painting tool. You can use it to edit the existing geometry as well as use it as a replacement for other tools/functions such as melting points, moving and duplicating elements, minor knife functions, extruding, tweaking, and so on.

Edge Cut
Hotkey: MF

This tool allows you to interactively subdivide the selected edges, see Fig. 26. This tool works in the **Edges** mode only. You can also use the **Shift**, **Ctrl**, and **Ctrl+Shift** keys to adjust scale, offset, and subdivision, respectively. You can also set these properties numerically by specifying values for the **Offset**, **Scale**, and **Subdivision** parameters in the **Attribute Manager | Options** tab. When the **Create N-gons** check box in the **Options** tab is selected, N-gons will be created else triangles will be created.

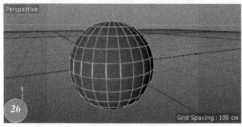 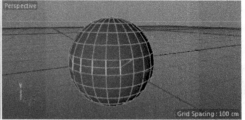

Line Cut
Hotkeys: MK, K

This tool has replaced the old **Knife** tool but if offers more features then the **Knife** tool. It can be used to cut polygon and spline objects. It works in all three modes: **Points, Edges,** and **Polygons.** There are many options available in the **Options** tab to control the behavior of the tool. To use this tool, click and drag to create a line or use multiple mouse clicks to create a line with multiple control points. The cut line will be projected onto the object to be cut. If you want to create additional point on the line, just click on it. The cut line will remain visible and you can edit it. Choose another tool or press **Esc** to complete the operation. You can choose the slice operation from the **Slice Mode** drop-down available in the **Line Cut** tool's properties in **Attribute Manager.** Fig. 27 shows split operation performed using the **Line Cut** tool.

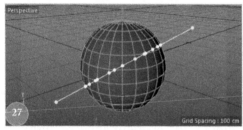 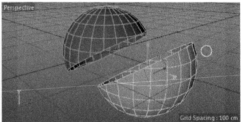

If you create a cut and switch to another view, cut points will be turned into the control points. The colored cut points can be grabbed and moved. The white cut points are automatically created when you cut the points. You can adjust the colored points any time. The following scheme is used for colors:

- **Yellow:** Control points lie freely in space
- **Blue:** Control points lie on a polygon
- **Red:** Control points lie on an edge
- **Green:** Control points lie on a vertex
- **White:** Points of the cut line with edges

You can use the following hotkeys with this tool:

- **Shift+Click** on a newly created point to move a control point along the a straight section.
- **Ctrl+Click** on a point to delete it.
- **Shift+Drag** the start or end point to take the **Angle Constrain** setting into effect. The rotation will be quantized.
- **Shift+Ctrl** drag a point; the neighboring cut points will not move.

- You can also cut an object using a spline that partially overlaps the object that you want to cut. Pressing **Ctrl** and then clicking on the spline with project the line onto the object and cut it.

Plane Cut
Hotkeys: KJ, MJ

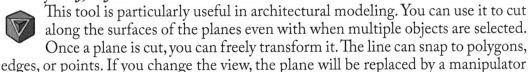

This tool is particularly useful in architectural modeling. You can use it to cut along the surfaces of the planes even with when multiple objects are selected. Once a plane is cut, you can freely transform it. The line can snap to polygons, edges, or points. If you change the view, the plane will be replaced by a manipulator [see Fig. 28]. You can use this manipulator to freely transform the plane.

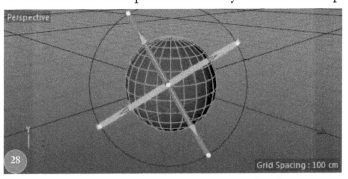

Loop/Path Cut
Hotkeys: KL, ML

This tool is primarily used to subdivide the edge loops interactively. It supports two modes: automatic loop selection [**Mode=Loop**] and manually created loop selection [**Mode=Path**]. Whenever you click on an edge, the HUD element will be activated in the viewport. The slider in HUD represents the length of the edge that you have clicked. It is represented by a green line in the viewport.

Ctrl+click on the slider to create a new loop cut. If you drag a handle with **Ctrl** held down, the corresponding loop cut will be duplicated. If you move the slider with **Shift** held down, the **Quantize Step** settings will be considered and handle will be snapped to fixed locations on the slider. The following hotkeys work in the viewport:

- **Ctrl:** Duplicates the cut.
- **Shift:** Uses the **Quantize Step** settings.
- **MMB:** Changes the cut count.
- **Ctrl+Shift+LMB:** Drag to left or right to control the bidirectional cut length.

Bevel
Hotkey: MS

The **Bevel** tool is a power tool in CINEMA 4D. You can use this tool to change harsh edges and corners into flattened, rounded, and soft elements, see Fig. 29. This tool behaves differently in each mode.

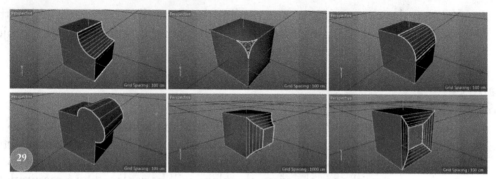

Bridge
Hotkeys: MB, B

 This tool works in all three modes and allows you to create connections between the unconnected surfaces, see Fig. 30.

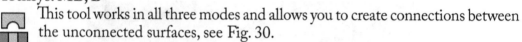

⚠ *Caution: Bridging Polygons*
In the Polygons mode, first you need to select the polygons you want to connect.

Weld
Hotkey: MQ

You can use this tool to weld several points on a polygon object or spline points to a single point. To weld points, select them and then using the **Weld** tool click on a point where the welded point will be placed.

Stitch and Sew
Hotkey: MP

You can use this tool to join edges with same number of points. It works in all three modes. However, it has some limitations in the **Polygons** mode. In this mode, most of the times, a polygon selection is required.

Close Polygon Hole
Hotkey: MD

This tool is used to fill a hole in a polygon mesh.

Extrude
Hotkeys: MT, D

This tool is used to extrude the selected points, edges, or polygons. If no elements are selected, it extrude the whole object. You can also interactively extrude in the viewport by dragging the mouse pointer to the left or right, see Fig. 31.

Extrude Inner
Hotkeys: M~W, I

The functioning of this tool is similar to that of the **Extrude** tool, however, you can extrude polygons inwards or outwards. The object shown in Fig. 32 is created using a combination of the **Extrude** and **Extrude Inner** tools.

Matrix Extrude
Hotkeys: MX

The functioning of the **Matrix Extrude** tool is similar to the **Extrude** tool, but with one difference: you can make as many extrusion steps as you want in one step, see Fig. 33. You can define the values for move, rotation, and size from the **Options** tab. You can also use **Shift**, **Ctrl**, and **Alt** to interactively set size, rotation, and directions values, respectively.

Smooth Shift
Hotkey: MY

The **Smooth Shift** tool works similar to the **Extrude** tool. However, the value specified for the **Maximum Angle** value will be used to determine if new connecting surface should be created between polygons. Fig. 34 shows the initial selection of the polygons, and extruded polygons using **Maximum Angle** value of **71**, respectively.

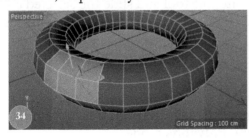 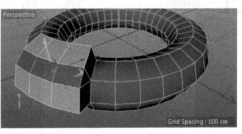

Transform Tools Sub-menu

The following commands are available in the **Transform Tools** sub-menu:

Brush
Hotkey: MC

 This tool works in all three modes. This tool allows you to deform polygon mesh and paint/edit vertex maps. When you select this tool a sphere of influence appears in the editor view. All points inside the sphere will be affected by the tool. You can interactively change the size of the sphere by MMB dragging in the editor view to the left or to the right. To change strength, MMB drag vertically.

Iron
Hotkey: MG

 This tool behaves like a virtual iron and lets you smooth the uneven surfaces. The strength of the iron is controlled using the **Percent** value in the **Options** tab.

Magnet
Hotkey: MI

You can use this tool to pull sections out of polygons or spline objects.

Mirror
Hotkey: MH

You can use this tool to mirror points and polygons. It can also be applied on the splines. It works only in the **Points** and **Polygons** modes. You can also define mirroring axis interactively. To define the axis, click and drag in the editor view.

Slide
Hotkey: MO

This tool is used to move selected edges, and edge loops vertically outwards or inwards. To offset the edges, activate this tool and slide the selected edges. You can use the **Ctrl** key to create copy of selected edges. The **Shift** key is used to push edges inwards or outwards.

Normal Move
Hotkey: MZ

 This interactive tool allows you to move the selected polygons in the direction of their normals. It works only in the **Polygons** mode.

Normal Scale
Hotkey: M#

 This interactive tool allows you to scale the selected polygons in the direction of their normals. It works only in the **Polygons** mode.

Normal Rotate
Hotkey: M

 This interactive tool allows you to rotate the selected polygons in the direction of their normals. It works only in the **Polygons** mode.

Weight Subdivision Surface
Hotkey: MR

 This tool is used with the **Subdivision Surface** objects. You can use it to weight subdivision surfaces.

N-Gons Sub-menu

The following commands are available in the **N-Gons** sub-menu:

N-gon Triangulation
Hotkey: UT

 While rendering and animating n-gons, CINEMA 4D triangulates them internally. If you want to preview them in the viewport, select the **N-gon Lines** check box from the **Viewport Settings | Filter** tab. The **N-gon Triangulation** command is a toggle switch. When on, an n-gon will be internally re-triangulated each time you move any n-gon's points. When off, you can triangulate the n-gons manually by using command available in the **Mesh | N-gons** sub-menu.

Retriangulate N-gons
Hotkey: UG

 See the **N-gon Triangulation** description.

Remove N-gons
Hotkey: UE

 This command can be used to convert the selected objects' n-gons to triangles and quadrangles.

Normals Sub-menu

The following commands are available in the **Normals** sub-menu:

Align Normals
Hotkey: UA

 You can use this command to adjust and re-align the incorrect surface normals to the correct direction. The normals are used by CINEMA 4D to recognize an object's inner and outer surfaces.

Reverse Normals
Hotkey: UR

 You can use this command to reverse the normals of the object.

Break Phong Shading

 Use this command to break the phong shading.

Unbreak Phong Shading

 You can use this command to restore the phong shading.

Select Broken Phong Edges

 This command selects all broken phong edges.

Axis Center Sub-menu

The following commands are available in the **Axis Center** sub-menu:

Axis Center

 This tool is used to quickly specify the object axis of a polygonal object. Choose this tool to open the **Axis Center** dialog box. Set options in this dialog and then click **Execute** to set the object axis.

Center Axis to

 The object axis will be moved but no geometry.

Center Object to

 The object geometry will be moved along the axis.

Center Parent to

 The object axis of the parent object of the object selected in **Object Manager** and its corresponding geometry will be moved onto the selected object.

Center to Parent

 The selected object's axis and its corresponding geometry will be placed onto the parent object.

View Center

The object axis and the corresponding geometry will be moved to the center of the view.

Working with Modeling Objects

CINEMA 4D offers several modeling functions that provide some special modeling features. You can access them from the **Modeling** sub-menu of the **Create** menu. You can also access them from the **Modeling** command group of the **Standard** palette, see Fig. 35. The function available in the **Modeling** command group and related functions are discussed next:

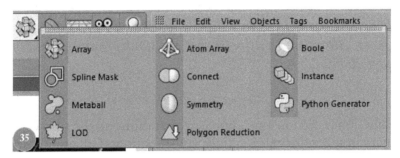

Array

This function is used to create copies of an object. The copied objects can be arranged in a spherical or wave form and are placed around the origin of the array object, see Fig. 36. The amplitude of the wave can be animated. The object you want to copy must be child of the array.

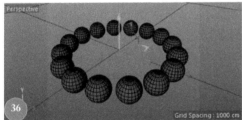
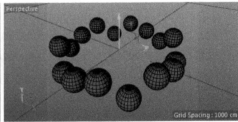

Atom Array

You can use the **Atom Array** function to create atomic lattice structure from the child objects. When you apply this function, all edges are replaced with the cylinders and points are replaced with the spheres, see Fig. 37.

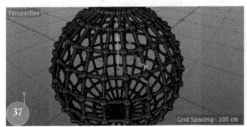
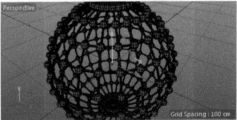

Boole

This function is used to apply boolean operations on the primitives or polygons, see Fig. 38. You can also use this function on hierarchies. The two objects on which you want to apply this function should be children of the **Boole** object. The default operation for this function is A subtract B. Therefore, the order of the child objects is important.

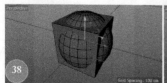  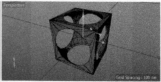

 Note: Boolean Operations
You need to make sure the objects to which you want to apply this function should have closed volume and cleanly structured otherwise unwanted results may occur. Also, note that the higher the subdivisions of the objects, cleaner the cut will be.

Spline Mask

This function is used to apply boolean operations on the splines. This function produces best results when splines are smooth and all are in the same plane, see Fig. 39.

 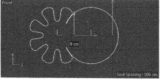 

Tip: Spline in Polygons mode
*You can display a closed booled spline shape in the **Polygons** mode, see the right image in Figure F39, by making it editable. To this to work, you need to select the **Create Cap** check box from the **Spline Mask's Object** tab.*

Connect

This function combines separate objects using a defined tolerance. It also gives you ability to weld them together, see Fig. 40. To use this function, select all the objects that you want to connect and then press **Alt+G** to group them under a **Null** object and then make **Null** child of the **Connect Object**. To smooth the connections, you can use a **Subdivision Surface** object.

Instance

You can use this function to create instances of an object. An instance does not have its own geometry and it takes far less memory than the copied geometries. However, **MoGraph's Cloner Object** is much more powerful than the **Instance** Object.

Metaball

This function creates an elastic skin over the spline and objects points. You can use parametric objects, splines, and polygon objects with this function, see Figs. 41 and 42.

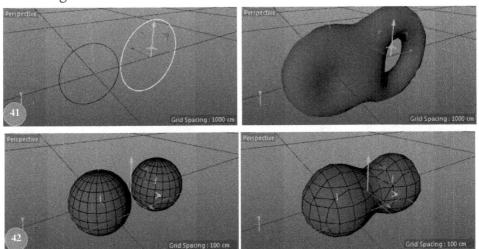

Symmetry Object

If you are creating a model that is symmetric in nature, you only have to model half the model. The other half you can generate using this function. Only the part on which you applied this function will have the points. If you manipulate these points, the action will be reflected in the other part as well. This object works with geometry only, not with lights, camera, and so on.

Python Generator

This function is used to enter **Python** code which can be used to generate the geometry.

LOD

Some scenes can be very complex when they contain many objects with details. It is very rare that a camera would be able to depict all details of the scene either because the objects are not in the field of view of the camera or they are far away in the scene that the details can't be seen. In such cases, the LOD object helps in speeding the workflow in the viewport or for rendering.

If you want ab object to be affected by an LOD object, you should make it child of the LOD object. Fig. 43 shows a **Cloner** object which is child of the **LOD** object. Notice in the **Attribute Manager | LOD | Object | Level -0- Cloner** area, the display mode is set to **Lines** and it is only affecting the **Cloner** object in the scene.

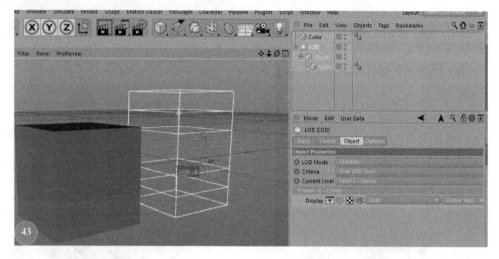

Polygon Reduction

This tool is very useful in optimizing the models with the large number of polygons such as 3D scanned objects. It is often difficult to edit such models. This tool allows you to reduce polygon count of such objects by maintaining the original shape of the object to the highest degree. The object should be the child of the **Polygon Reduction** tool.

Fig. 44 shows a model before [Total polygons=11320] and after [Resulting polygons=4541] applying the **Polygon Reduction** tool. A value of **80%** used for the **Reduction Strength** attribute.

Exploring Deformers

The deformer objects in CINEMA 4D are used primarily to deform the shape of the primitive objects, Generator objects, polygon objects and splines, see Fig. 45. Unlike the objects that belong to **Generators** and **Modeling** command groups, deformers act as children of their parents. A deformer will have no effect on the geometry if it is at the top of the hierarchy. It will affect its parent object as well as the hierarchy below that parent.

To add a deformer, select the parent object in **Object Manager**, hold **Shift** and then choose the desired deformer from the **Standard** palette | **Deformer** command group, see Fig. 46. You can apply multiple deformers on the same object.

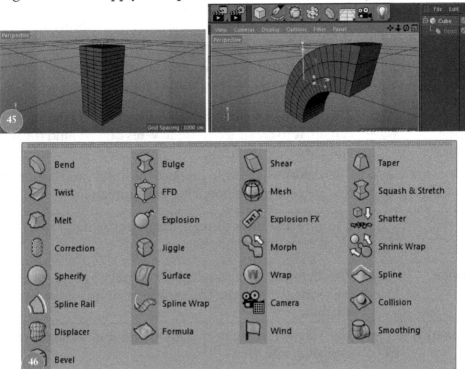

Caution: Deformer Object

*The **Deformer** object does not work in conjunction with the following functions in CINEMA 4D: **Explosion Object**, **ExplosionFX**, **Polygon Reduction**, **Spline Deformer**; and **Shatter Object**.*

The following table summarizes the deformer objects available in CINEMA 4D.

Table 1: The deformer objects available in the CINEMA 4D		
Function	**Icon**	**Description**
Bend		This deformer bends an object. You can drag the orange handle on the deformer surface to interactively control the amount of bend.
Bulge		Use this deformer to make an object bulge or contract. You can drag the orange handle on the deformer surface to interactively control the amount of bulge.
Shear		It shears an object.
Taper		It tapers [narrows or widens towards on end] an object.

Table 1: The deformer objects available in the CINEMA 4D

Function	Icon	Description
Twist		It twists an object around its Y-axis. For smooth twist, ensure there are sufficient number of subdivisions along the twist axis.
FFD		This deformer deforms objects using a grid points. This deformer works in the **Points** mode only.
Mesh Deformer		This deformer somewhat works like the **FFD** deformer. You can use it to create a custom low-res cage around the model and then deform freely.
Squash and Stretch		This deformer allows to produce the squash and stretch effect that you see in the bouncing ball animation.
Melt		Use this deformer to melt the object radially from origin [Y plane] of the deformer.
Explosion		This deformer lets you explode an object to its constituent polygons. To animate the explosion, animate the **Strength** attribute.
ExplosionFX		Use this deformer to quickly create and animate realistic explosion effects.
Shatter		Use this deformer to shatter objects into individual polygons which then fall to the ground plane.
Correction		This deformer allows you to access the points in their deformed state and then allows you to change the position of these points.
Jiggle		You can use this deformer to create secondary motion for a character's motion.
Morph		This deformer lets you blend in the morph targets within the region of influence of the deformer.
Shrink Wrap		This deformer allows you to shrink wrap the source object onto the target object.
Spherify		You can use this deformer to deform an object into a spherical shape.
Surface		Use this deformer to make an object follow the surface deformations of another object. You can use this tag for example to quickly attach stitches to a cloth that is being deformed by a **Cloth** tag.

Table 1: The deformer objects available in the CINEMA 4D

Function	Icon	Description
Wrap		You can use this deformer to wrap flat surface onto a curved surface.
Spline Deformer		This deformer takes two splines: original spline and modifying spline. It considers the difference in the position and shape of the two splines and then deforms the object accordingly.
Spline Rail		It deforms the polygon objects using upto four splines. These splines define the target shape.
Spline Wrap		This deformer allows you to deform an object along a spline.
Camera Deformer		You can use this deformer to deform an object based on the grid overplayed on the camera view.
Collision Deformer		This deformer deforms the objects using the collision interaction. You can think it as if a soft surface being pulled or pushed when it collides with another surface.
Displacer		This deformer allows you to create effects like created by the displacement mapping.
Formula		You can use this deformer and a mathematical formula to deform objects.
Wind Deform		Use this deformer to create waves on an object. The effect of this deformer will be along the deformer's positive X direction.
Smoothing		You can use this deformer to smooth a surface. It can make a surface behave almost like a cloth. It adds lots of flexibility to the modeling workflow when trying to model cloth objects.
Bevel		The functioning of this deformer is almost similar to the **Bevel** tool.

Tutorials

Before you start the tutorials, create a folder with the name **chapter-m4**. We'll use this folder to host all the tutorial files and other resources.

Tutorial 1: Creating a Circular Hole in the Geometry

In this tutorial, we will create a circular hole in the geometry [Fig. T1]. The following table summarizes the tutorial:

Table T1	
Flow: The following sequence will be used to create the model: **1.** Create a **Polygon** object. **2.** Create a hole in the **Polygon** object using the polygon modeling tools.	
Difficulty level	Beginner
Estimated time to complete	10 Minutes
Topics	• Getting Started • Creating the Hole
Resources folder	**chapter-m4**
Tutorial units	**Centimeters**
Final tutorial file	**m4-tut1-finish.c4d**

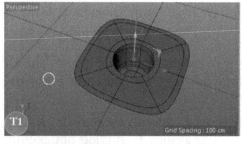 

Getting Started

Start a new scene in CINEMA 4D and set units to **Centimeters**.

Creating the Hole

Follow the steps given next:

1. Click **Polygon** from the **Standard** palette | **Object** command group to create a polygon in the editor view. In the **Attribute Manager | Polygon Object | Object** tab, change **Segments** to **2**. Press **C** to make object editable. Press **NB** to enable the **Gouraud Shading (Lines)** mode.

2. Activate the **Points** mode. Select the four corner points and the center point. Press **MM** to connect the points and create edges [Fig. T2].

What Next?
Now, we will bevel the center vertex to create a polygon. Then, we will delete the center polygon to create the hole.

3. Select the center point. Press **MS** to invoke the **Bevel** tool and then bevel the selected edges. On the **Attribute Editor | Bevel | Tool Option** tab, change **Offset** to **22** [Fig. T3].

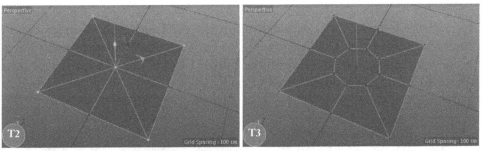

4. Activate the **Polygons** mode and then select the center polygon. Delete the polygon [Fig. T4]. Activate the **Edges** mode and then select the boundary loop [Fig. T5] and then extrude it down by using the **Ctrl** key and **Move** tool [Fig. T6].

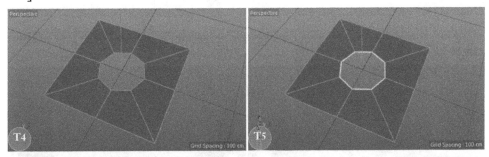

5. Using the **Ctrl** key and **Scale** tool, create two inner extrusions [Fig. T7]. Activate the **Points** mode and then select point loop using the **Loop Selection** tool [Fig. T8]. Press **MQ** to activate the **Weld** tool and then weld the vertices at the center [Fig. T9].

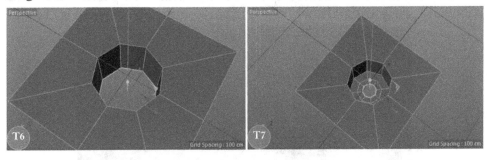

What Next?
Now, we will select the edge loops and then we will bevel and chamfer them.

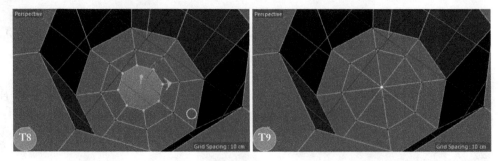

6. Activate the **Edges** mode, press **UL** to invoke the **Loop Selection** tool and then select the loops [Fig. T10]. Press **MS** to invoke the **Bevel** tool and then bevel the selected edges. On the **Attribute Editor | Bevel | Tool Option** tab, change **Offset** to **1** and **Subdivision** to **1** [Fig. T11].

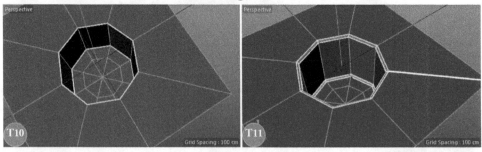

7. Activate the **Edges** mode and then select edges [Fig. T12]. Press **UZ** to melt the edges [Fig. T13].

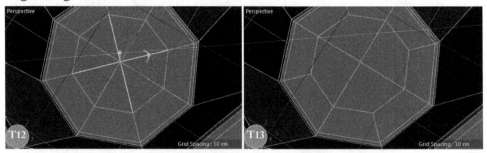

8. Choose **Subdivision Surface** from the **Standard** palette | **Generators** command group with **Alt** held down to smooth the object [see Fig. T14].

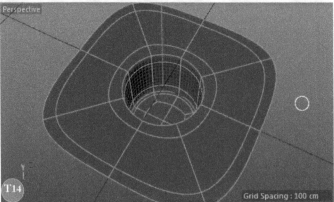

Tutorial 2: Creating a Cylinder with Holes

In this tutorial, we will create a cylinder with holes [Fig. T1]. The following table summarizes the tutorial:

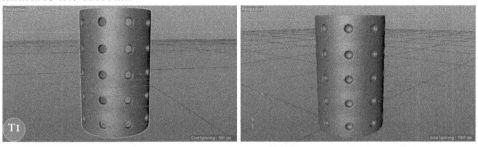

Table T2	
Flow: The following sequence will be used to create the model: **1.** Create a **Polygon** object. **2.** Create a hole in the **Polygon** object using the polygon modeling tools. 3. Create a strip of the polygon objects using the **Duplicate** command. **3.** Use the **Bend** deformer to create shape of the cylinder. **4.** Smooth the object using the **Subdivision Surface** generator.	
Difficulty level	Beginner
Estimated time to complete	20 Minutes
Topics	• Getting Started • Creating the Model
Resources folder	**chapter-m4**
Tutorial units	**Centimeters**
Final tutorial file	**m4-tut2-finish.c4d**

Getting Started

Start a new scene in CINEMA 4D and set units to **Centimeters**.

Creating the Model

Follow the steps given next:

1. Follow the steps 1 through 4 of the **Creating the Hole** section of the Tutorial 1.

2. Activate the **Edges** mode and then press **UL** to invoke the **Loop Selection** tool and then select the edge loop [Fig. T2].

3. Press **MT** to invoke the **Extrude** tool and then extrude the loop inward [Fig. T3].

4. Activate the **Model** mode and make sure **Polygon** is selected in **Object Manager**. Choose **Arrange Objects | Duplicate** from the **Tools** menu. On the **Attribute Manager | Options** tab, change **Mode** to **Linear**. On the **Position** section, clear

the **Y** and **Z** check boxes and then enter **100** in the **Move X** field. This action creates **8** copies of the **Polygon** object [Fig. T4].

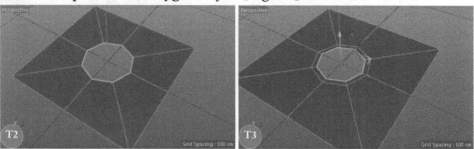

5. Select everything in **Object Manager** and then right-click and choose **Connect Objects + Delete** from the popup menu. Activate the **Points** mode, select all points, and then press **UO** to optimize the object.

What just happened?

*Here, we have executed the **Connect Objects + Delete** command. This command connects the selected objects, create a new surface, and delete the selected objects.*

6. Activate the **Model** mode and invoke the **Move** tool. Press **L** to enable the axis. Press **Shift+S** to enable snap and make sure **Vertex Snap** is on from the **Snap** menu. Snap the axis, as shown in Fig. T4. Press **L** to disable the axis.

What next?

*Now, we will create few more strips and then we will use the **Bend** deformer to create the shape of the cylinder.*

7. Create a duplicate of the object, change its axis [refer to Fig. T5], as discussed in the previous step, and then align it [refer to Fig. T5]. Similarly, create few more strips [Fig. T6].

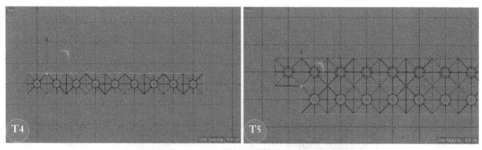

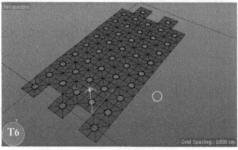

8. Select everything in **Object Manager** and then right-click and choose **Connect Objects + Delete** from the popup menu. Activate the **Points** mode, select all points, and then press **UO** to optimize the object.

9. Activate the **Model** mode and then rotate the object by **90** degrees [Fig. T7].

10. Choose **Bend** from the **Standard** palette | **Deformer** command group with **Shift** held down to attach a bend modifier to the object. On the **Attribute Manager** | **Bend Object** | **Coord** tab, change **R.B** to **90**. On the **Object** tab, change **Size** fields to **5, 1000,** and **500,** respectively. Change **Strength** to **400** [see Fig. T8].

11. Select everything in **Object Manager** and then right-click and choose **Connect Objects + Delete** from the popup menu. Activate the **Points** mode, select all points, and then press **U** and then **Shift+O** to open the **1** dialog. In this dialog, set **Tolerance** to **0.03** and click **OK** to optimize the object.

12. Activate the **Polygons** mode and then select all polygons. Press **MT** to enable the **Extrude** tool and then extrude them outward by **8** units [Fig. T9].

13. Select the top and bottom edge loops of the cylinder [Fig. T10]. Press **MS** to invoke the **Bevel** tool and then bevel the selected edges. On the **Attribute Editor** | **Bevel** | **Tool Option** tab, change **Offset** to **1** and **Subdivision** to **1**. On the **Topology** tab, change **Mitering** to **Uniform** [Fig. T11].

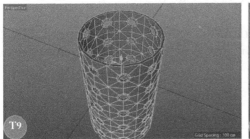

14. Choose **Subdivision Surface** from the **Standard** palette | **Generators** command group with **Alt** held down to smooth the object [see Fig. T12].

15. If you want to close the holes, activate the **Edges** mode and then invoke the **Close Polygon Hole** by pressing **MD**. Now, click on the border loops to close the holes [see Figs. T13 and T14].

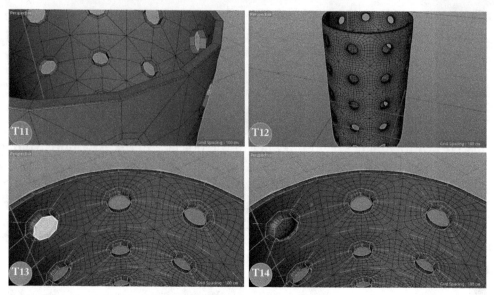

Tutorial 3: Creating a Solid Model

In this tutorial, we will create a solid model [Fig. T1].

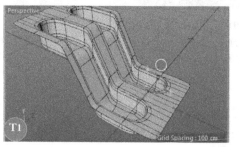 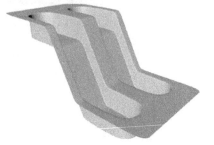

The following table summarizes the tutorial:

Table T3	
Flow: The following sequence will be used to create the model:	
1. Create a **Plane** object and then extrude it to create the basic shape. **2.** Use the **Polygon Pen** tool to create the circular arcs. **3.** Use the **Subdivision Surface** generator to smooth the object.	
Difficulty level	Beginner
Estimated time to complete	20 Minutes
Topics	• Getting Started • Creating the Model
Resources folder	**chapter-m4**
Tutorial units	**Centimeters**
Final tutorial file	**m4-tut3-finish.c4d**

Getting Started

Start a new scene in CINEMA 4D and set units to **Centimeters**.

Creating the Model

Follow the steps given next:

1. Click **Plane** from the **Standard** palette | **Object** command group to create a plane in the editor view. In **Attribute Manager** | **Plane** | **Object** tab, change **Width** to **150**, **Height** to **200**, **Width Segments** to **1**, and **Height Segments** to **1**. Press **NB** to enable the **Gouraud Shading (Lines)** mode.

2. Press **C** to make the plane editable.

3. Activate the **Edges** mode and then select the edge [Fig. T2]. Press **Ctrl** and drag the edge to extrude it [Fig. T3]. Extrude again [Fig. T4].

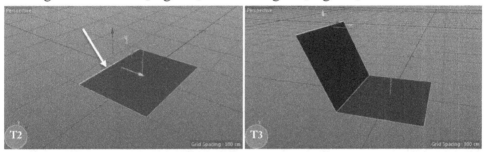

4. Press **UB** to invoke the **Ring Selection** tool and then select the edge ring [Fig. T5]. Press **MF** to invoke the **Edge Cut** tool. On the **Attribute Editor | Edge Cut | Options** tab, clear the **Create N-gons** check box and then change **Subdivision** to **4**. Now, **Shift**-drag to create four edges [Fig T6.].

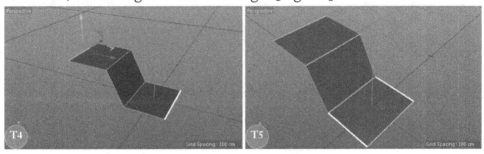

5. Select the edges [Fig. T7] using the **Move** tool and then scale them in using the **Scale** tool [Fig. T8]. Make a selection using the **Ring Selection** tool [see Fig. T9]. Now, press **MM** to connect the edges [Fig. T10].

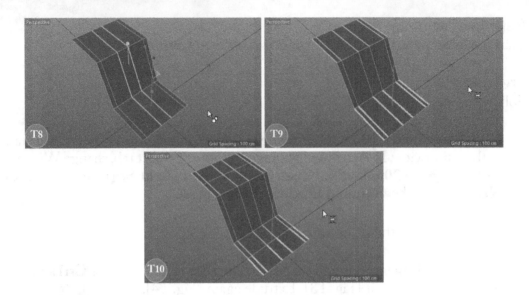

> **What next?**
> Now, we will use the Polygon Pen tool to create arc from the edges. These arc will form the rounder corners.

6. Press **ME** to invoke the **Polygon Pen** tool and then on **Attribute Editor,** select the **Create Semi Circle** check box. Now, press **Ctrl+Shift** and hover the mouse of the edge [Fig. T11]. When you see a semi circle, click to create the semi circle. Don't release the mouse button and then drag until tooltip shows **Sub:4** [Fig. T12].

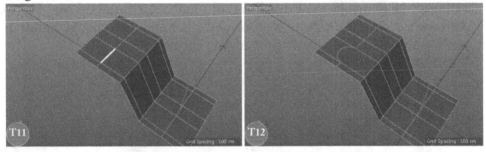

7. Similarly, create other semi circles [Fig, T13]. Notice that by creating the semi-circles, we have created n-gons. Let's convert them to quads.

8. Make sure the **Polygon Pen** tool is active. Click on a point on the circle and then click on the outer edge of the model to create an edge [Fig. T14]. Similarly, connect all four semi-circles.

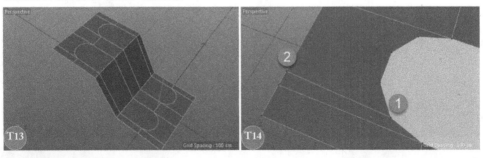

9. Select the polygons [Fig. T15] and then **Ctrl-drag** them to create the extrusion along the -Y axis [Fig. T16].

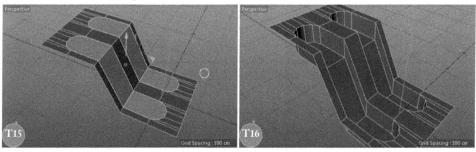

10. Press **UL** to invoke the **Loop Selection** tool and then select loops [Fig. T17]. Press **MS** to invoke the **Bevel** tool and then bevel the selected edges. On the **Attribute Editor | Bevel | Tool Option** tab, change **Offset** to **1.5** and **Subdivision** to **1**. On the **Topology** tab, change **Mitering** to **Uniform** [Fig. T18].

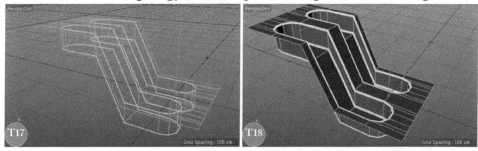

11. Press **KL** to invoke the **Loop/Path Cut** tool and ensure that **Mode** is set to **Loop** in **Attribute Manager**. Now, create two cuts [Fig. T19].

12. Choose **Subdivision Surface** 🔲 from **Standard Palette | Generator** command group with **Alt** held down to smooth the object [see Fig. T20].

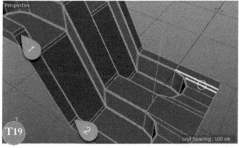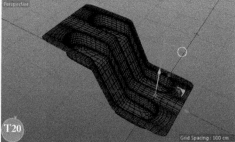

Tutorial 4: Creating a Solid Model

In this tutorial, we will create a solid model [Fig. T1]. The following table summarizes the tutorial:

Table T4	
Flow: The following sequence will be used to create the model: **1.** Create a **Tube** object. **2.** Create one section of the model using polygon editing tools. **3.** Complete the shape using the **Array** object. **4.** Connect all geometries. **5.** Smooth the object using the **Subdivision Surface** generator.	
Difficulty level	Beginner
Estimated time to complete	20 Minutes
Topics	• Getting Started • Creating the Model
Resources folder	**chapter-m4**
Tutorial units	**Centimeters**
Final tutorial file	**m4-tut4-finish.c4d**

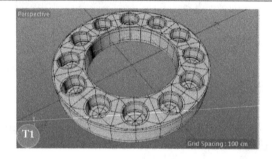 

Getting Started

Start a new scene in CINEMA 4D and set units to **Centimeters**.

Creating the Model

Follow the steps given next:

1. Click **Tube** from **Standard** palette | **Object** command group to create a tube in the editor view. In the **Attribute Manager | Tube Object | Object** tab, change **Inner Radius** to **120**, **Outer Radius** to **200**, **Cap Segments** to **2**, **Rotation Segments** to **24**, and **Height** to **50**.

2. Press **C** to make object editable. Press **NB** to enable the **Gouraud Shading (Lines)** mode.

3. Activate the **Points** mode, select all points, and then press **UO** to optimize the object [Fig. T2].

⊕ ***What next?***
*We will first extract a part of the **Tube** object and then create a hole in the extracted part. Next, we will use the **Array** tool to create the complete circular geometry.*

4. Activate the **Polygons** mode and then select the polygons [Fig. T3]. Press **UI** to invert the selection. Press **Delete** to delete the selected polygons [Fig. T4] and then press **UO** to optimize the object.

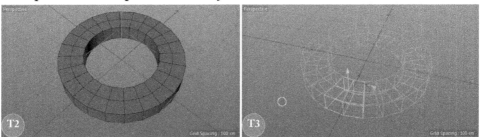

5. Activate the **Points** mode and then select the middle point [Fig. T5]. Press **MS** to invoke the **Bevel** tool and then bevel the selected point. On the **Attribute Editor | Bevel | Tool Option** tab, change **Offset** to **29** [Fig. T6].

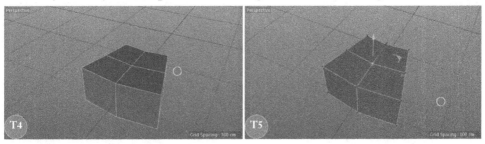

6. Activate the **Edges** mode and then select edges [Fig. T7]. Now, press **MM** to connect the edges [Fig. T8]. Similarly, connect other edges [Fig. T9].

7. Activate the **Points** mode and then select the points [Fig. T10]. Using the **Scale** tool, uniformly scale the points to create a circular shape [Fig. T11].

8. Press **ME** to invoke the **Polygon Pen** tool and then connect the points to get rid of the nGons and create quads [Fig. T12]. Activate the **Polygons** mode and then select and delete the polygons created using the **Bevel** operation [Fig. T13].

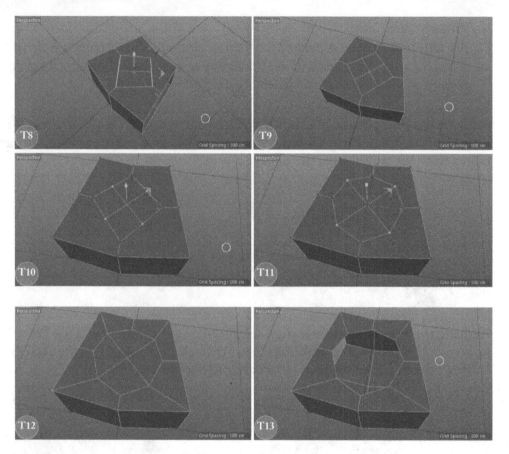

9. Activate the **Edges** mode and then press **UL** to invoke the **Loop Selection** tool. Select the loop [Fig. T14]. Using the **Move** tool and **Ctrl**, extrude the loop downward [Fig. T15].

10. Using the **Scale** tool and **Ctrl**, extrude the loop inwards twice [Fig. T16]. Activate the **Points** mode and then select the points [Fig. T17]. Press **MQ** to activate the **Weld** tool and then weld the vertices at the center [Fig. T18].

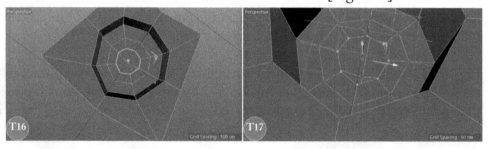

11. Press **UL** to invoke the **Loop Selection** tool and then select the loops [Fig. 19]. Press **MS** to invoke the **Bevel** tool and then bevel the selected edges. On the **Attribute Editor | Bevel | Tool Option** tab, change **Offset** to 1 and **Subdivision** to 1 [Fig. 20].

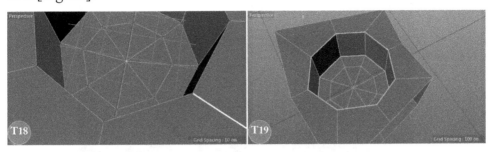

12. Choose **Array** from the **Standard** palette | **Modeling** command group with **Alt** held down. On the **Attribute Editor | Array Object | Object** tab, change **Radius** to **0** and **Copies** to **11** [Fig. T21].

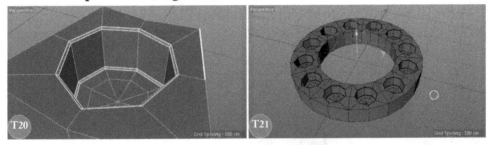

13. On **Object Manager**, select **Array** and then press **C**. This action will create **12** copies of the tube object and also group then under a **Null** object. On **Object Manager**, expand **Null** and select all objects. Now, right-click and then choose **Connect Objects + Delete** from the popup menu and create single object. Drag combined object out of **Null** and then delete **Null**.

14. Activate the **Points** mode and then select all points. Press **UO** to optimize the model. Activate the **Edges** mode and then select the boundary loops [Fig. T22].

15. Press **MS** to invoke the **Bevel** tool and then bevel the selected edges. On the **Attribute Editor | Bevel | Tool Option** tab, change **Offset** to 1 and **Subdivision** to 1. On the **Topology** tab, change **Mitering** to **Uniform** [Fig. T23].

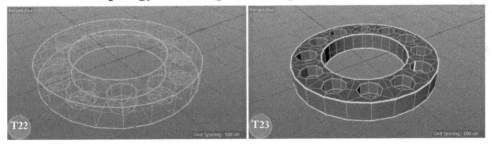

16. Switch to the **Model** mode. Choose **Subdivision Surface** ⬡ from the **Standard** palette | **Generators** command group with **Alt** held down to smooth the object [see Fig. T24].

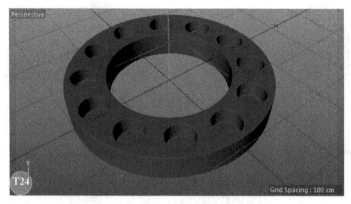

Tutorial 5: Creating a Solid Model

In this tutorial, we will create a solid model [Fig. T1].

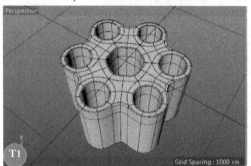 

The following table summarizes the tutorial:

Table T5	
Flow: The following sequence will be used to create the model: 1. Create a **Tube** object. 2. Use the **Array** command to create copies. 3. Use the **Polygon Pen** tool to connect the objects. 4. Extrude and bevel the edges. 5. Smooth the object using the **Subdivision Surface** generator.	
Difficulty level	Beginner
Estimated time to complete	20 Minutes
Topics	• Getting Started • Creating the Model
Resources folder	**chapter-m4**
Tutorial units	**Centimeters**
Final tutorial file	**m4-tut5-finish.c4d**

Getting Started
Start a new scene in CINEMA 4D and set units to **Centimeters**.

Creating the Model
Follow the steps given next:

1. Click **Tube** from **Standard** palette | **Object** command group to create a tube in the editor view. In the **Attribute Manager | Tube | Object** tab, change **Inner Radius** to 140, **Outer Radius** to 200, **Rotation Segments** to 8, **Height** to 0, and **Cap Segments** to 1. Press **NB** to enable the **Gouraud Shading (Lines)** mode.

    $\rightarrow$ *What next?*
    *Now, we will use the **Array** tool to create 5 more copies of the Tube object. Then, we will use different tools to connect these shapes.*

2. Choose **Array** from **Standard** palette | **Modeling** command group with **Alt** held down. On the **Attribute Editor | Array Object | Object** tab, change **Radius** to 460 and **Copies** to 5 [Fig. T2].

3. Press **Ctrl+A** to select all objects in **Object Manager**. Press **C**. Select all objects under **Array** null and then right-click, choose **Connect Objects + Delete** from the popup menu.

4. Drag **Tube** out of the **Array** and delete the **Array** object.

5. Activate the **Edges** mode and then press **ME** to invoke the **Polygon Pen** tool. Extrude the edges and then snap them to opposites edges using Ctrl-drag operation [Figs. T3 and T4]. Similarly, make all connections [Fig. T5]. Activate the **Points** mode, select all points, and then press **UO** to optimize the object.

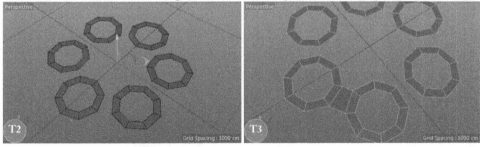

6. Activate the **Edges** mode and then select edges [Fig. T6]. Press **MM** to connect the edges [Fig. T7]. Similarly, make other connections [Fig. T8]. Activate the **Points** mode, select all points, and then press **UO** to optimize the object.

7. Select the points [Fig. 9] and then scale them in using the Scale tool [Fig. 10].

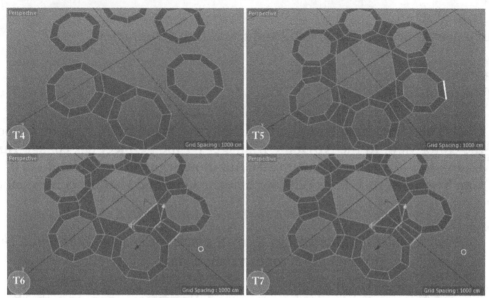

8. Activate the **Edges** mode and then select the edge loop [Fig. 11] and using the **Scale** tool and **Ctrl**, extrude the loop inwards [Fig. 12].

9. Select the edge loops [Fig. 13] and then extrude them down using the **Move** tool and **Ctrl** [Fig. 14].

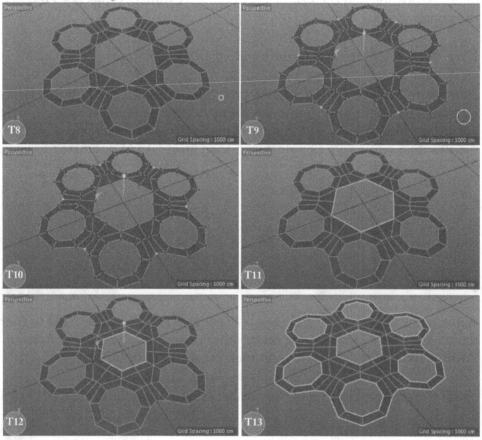

10. Select the edge loops [Fig. 15]. Press **MS** to invoke the **Bevel** tool and then bevel the selected edges. On the **Attribute Editor | Bevel | Tool Option** tab, change **Offset** to **5** and **Subdivision** to **1**. On the **Topology** tab, change **Mitering** to **Uniform** [Fig. 16].

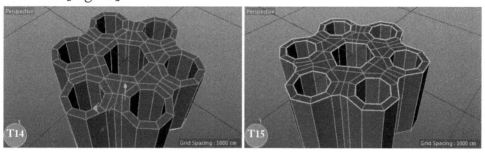

11. Choose **Subdivision Surface** 🗔 from **Standard** palette | **Generators** command group with **Alt** held down to smooth the object [see Fig. T17].

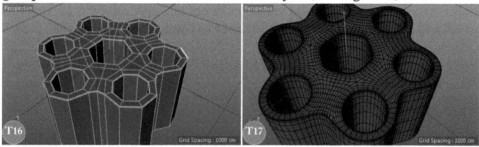

Tutorial 6: Creating a Serving Bowl

In this tutorial, we will create model of a bowl [Fig. T1].

The following table summarizes the tutorial:

Table T6
Flow: The following sequence will be used to create the serving bowl model: **1.** Create a **Cylinder** primitive. **2.** Edit object's elements to create shape using the **Extrude** tool. **3.** Use the **Bevel** tool to smooth the edges. **4.** Use the **Taper** deformer to create the final shape.

Difficulty level	Beginner
Estimated time to complete	20 Minutes
Topics	• Getting Started • Creating the Bowl

Table T6	
Resources folder	**chapter-m4**
Tutorial units	**Centimeters**
Final tutorial file	**m4-tut6-finish.c4d**

Getting Started

Start a new scene in CINEMA 4D and set units to **Centimeters**.

Creating the Bowl

Follow the steps given next:

1. Click **Cylinder** ⬚ from the **Standard** palette | **Object** command group to create a cylinder in the editor view. Rename **Cylinder** as **bowlGeo** in **Object Manager**. In the **Attribute Manager | bowlGeo | Object** tab, set **Radius** to **25.591**, **Height** to **13**, and **Height Segments** to **1**. Press **O** to frame the object in the editor view.

2. Press **NB** to enable the **Gouraud Shading (Lines)** mode, Fig. T2. Press **C** to make **bowlGeo** editable and then activate the **Points** mode. Select the top point,

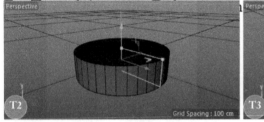
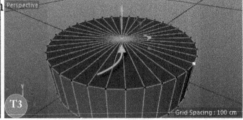

3. Select the bottom point, Fig. T4, and then press **UZ** to melt the point to get a plane surface, Fig. T5.

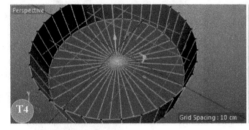
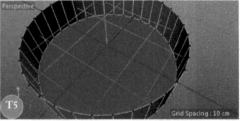

→ *What next?*
*If you move any of the bottom points using the **Move** ✛ tool, you would notice that these cap points are not connected with the rest of the geometry. Now, we will fix it.*

4. Undo the move operation, if any, and then click **Connect** ⬭ from the **Generator** tab with **Alt** held down to add make the **Connect** object parent of **bowlGeo**. In **Attribute Manager | Connect | Object** tab, select **Manual** from the **Phong Mode** drop-down. Ensure **Connect** is selected in **Object Manager** and then press **C** to convert it to a polygonal object.

What just happened?
The **Connect** *generator connects the points of the cap section to the rest of the geometry. Notice in* **Attribute Manager***, the* **Weld** *check box is selected. When on, CINEMA 4D welds the points using a tolerance value specified using the* **Tolerance** *parameter. Then, we converted the* **Connect** *object to an editable polygon object for farther changing the shape of the bowl.*

5. Now, activate the **Polygons** 🔲 mode. Select the bottom polygon, Fig. T6, and then **Ctrl+click** on **Tools** palette | **Points** 🔳 to select the bottom points, refer to Fig. T7. In **Coordinate Manager**, enter **40** in the **Size X** and **Size Z** fields to scale the points, Fig. T8.

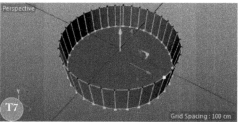

Note: Live Selection tool
Press and hold the **9** *key to temporarily enable the* **Live Selection** *tool.*

6. Activate the **Polygons** mode and then select the bottom polygon. Press **MT** to enable the **Extrude** 🔲 tool and then in the **Attribute Manager | Extrude | Options** tab, enter **1** in the **Offset** field to extrude the polygon, Fig. T9. You can also extrude polygon interactively in the editor view by dragging the mouse pointer to the left or right.

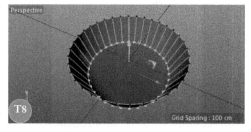

7. Press **MW** to enable the **Extrude Inner** 🔲 tool and then enter **1** in the **Attribute Manager | Extrude Inner | Options** tab | **Offset** field to extrude the polygon, Fig. T10. **Ctrl+click** on **Tools** palette | **Points** to select the points associated with the previously selected polygon. Press **UC** to collapse the points, Fig. T11. Activate the **Polygons** 🔲 mode and then press **Ctrl+A** to select all polygons.

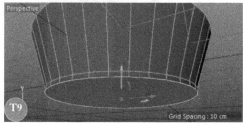
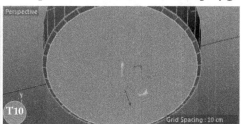

8. Press **MT** to enable the **Extrude** 🔲 tool and then in the **Attribute Manager | Extrude | Options** tab, enter **1** in the **Offset** field to give thickness to the bowl, Fig. T12. In **Object Manager**, rename **Connect** as **bowlGeo**.

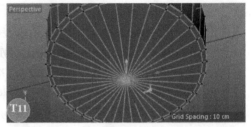

9. Press **UL** to select the **Loop Selection** tool 🔲 and then select the loops, in the **Edges** mode, as shown in Fig. T13. Press **MS** to activate the **Bevel** tool 🔲 and then in the **Attribute Manager | Bevel | Tool Option** tab, set **Offset** to **0.1** and **Subdivision** to **1** to chamfer the edges, Fig. T14. Similarly, chamfer the bottom and inner edges, Fig. T15.

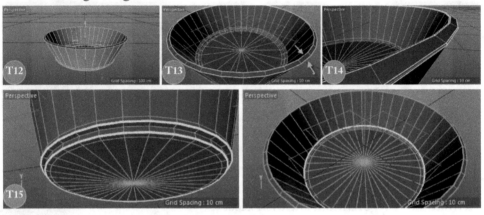

(?) What just happened?
I have added some edge loops at the top and bottom of the bowl. It will help us to retain the shape of the bowl when we will apply smoothing to it.

10. Activate the **Model** mode form the **Tools** palette. Now, choose **Taper** 🔲 from the **Deformer** tab with **Shift** held down to make the **Taper** object child of **bowlGeo**. In **Attribute Manager | Taper | Object** tab, set **Size X** to **25**, **Mode** to **Within Box**, and **Strength** to **-15%**, Fig. T16.

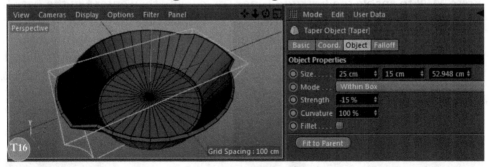

11. In **Object Manager**, create a copy of **Taper** object, Fig. T17 and then enter **90** in **Coordinate Manager | Rotation H** field, Fig. T18. Ensure **Taper.1** is selected in **Object Manager** and then in the **Attribute Manager | Taper | Object** tab, set **Size Z** to **70**, Fig. T19. Ensure **bowlGeo** is selected in **Object Manager** and

then choose **Subdivision Surface**  from the **Standard** palette | **Generators** command group with **Alt** held down to smooth the object.

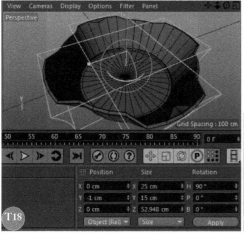

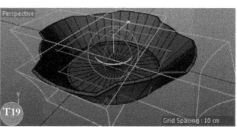

Tutorial 7: Creating a Kitchen Cabinet

In this tutorial, we will create model of a kitchen cabinet [Fig. T1].

The following table summarizes the tutorial:

Table T7	
Flow: The following sequence will be used to create the kitchen cabinet model: **1.** Use a **Cube** primitive to create the basic shape of the cabinet. **2.** Use the **Split**, **Extrude Inner** and **Extrude** functions to complete the model.	
Difficulty level	Beginner
Estimated time to complete	20 Minutes
Topics	• Getting Started • Creating the Bowl
Resources folder	**chapter-m4**

Table T7	
Tutorial units	**Centimeters**
Final tutorial file	**m4-tut7-finish.c4d**

Getting Started
Start a new scene in CINEMA 4D and set units to **Centimeters**.

Creating the Cabinet
Follow the steps given next:

1. Choose **Cube** 🔲 from the **Standard** palette | **Object** tab to create a cube in the editor view. Rename **Cube** as **cabinetGeo** in **Object Manager**. In the **Attribute Manager** | **cabinetGeo** | **Object** tab, set **Size X** to **38**, **Size Y** to **76**, and **Size Z** to **45**. Press **O** to frame the object in the editor view.

2. Press **NB** to enable the **Gouraud Shading (Lines)** mode, Fig. T2. Press **C** to make **cabinetGeo** editable. Select the top and bottom polygons of the **cabinetGeo** and then press **UP** to split the polygons.

   What just happened?
   *Here, I've applied the **Split** function on the selected polygons. As a result, CINEMA 4D creates a new object from the selected polygons leaving the original geometry unchanged, Fig. T3.*

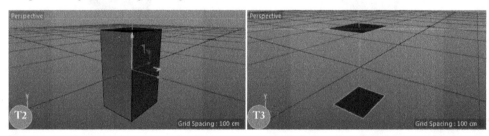

3. Rename the newly created object as **topbotGeo**. Select the two polygons of the **topbotGeo** and then press **MT** to enable the **Extrude** tool and then in the **Attribute Manager** | **Extrude** | **Options** tab, enter **3** in the **Offset** field to extrude the polygons, Fig. T4. Now, select the polygon, Fig. T5, and then move it by **3** units in the negative Z direction using the **Move** ✛ tool, Fig. T6.

4. Select the top polygon and then extrude the polygon by **5** units using the **Extrude** tool, Fig. T7. Now, select the polygon shown in Fig. T8 and extrude it by **5** units along the positive Z direction, Fig. T9.

5. Select the edges of **cabinetGeo**, refer Fig. T10 and then press **MM** to connect the edges, Fig. T10. Select the polygons, Fig. T11, and then press **MW** to activate the **Extrude Inner** 🔲 tool. In **Attribute Manager | Extrude Inner | Options** tab, clear the **Preserve Groups** check box and then set **Offset** to **2**, see Fig. T12.

6. Select the newly created polygons and then extrude them by **2.5** units, Fig. T13. Press **Ctrl+A** to select all points. Press **MS** to invoke the **Bevel** tool and then bevel the selected edges. On the **Attribute Editor | Bevel | Tool Option** tab, change **Offset** to **0.1** and **Subdivision** to **1**. On the **Topology** tab, change **Mitering** to

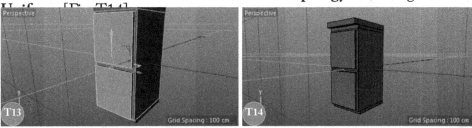

Tutorial 8: Creating a Book
In this tutorial, we will create model of a book [Figs. T1 and T2].

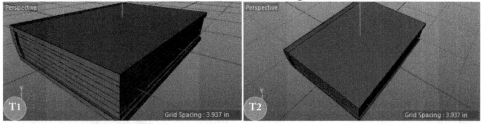

The following table summarizes the tutorial:

Table T8
Flow: The following sequence will be used to create the book:
1. Create a **Cube** primitive. **2.** Edit elements using the **Bevel** and **Extrude** tools to create final model.

Difficulty level	Beginner
Estimated time to complete	20 Minutes

Table T8	
Topics	• Getting Started • Creating the Book
Resources folder	**chapter-m4**
Tutorial units	**Inches**
Final tutorial file	**m4-tut8-finish.c4d**

Getting Started
Start a new scene in CINEMA 4D and set units to **Inches**.

Creating the Book
Follow the steps given next:

1. Choose **Cube** from the **Standard** palette | **Object** command group to create a cube in the editor view. Rename **Cube** as **bookGeo** in **Object Manager**. In the **Attribute Manager | bookGeo | Object** tab, set **Size X** to **7.44**, **Size Y** to **2**, and **Size Z** to **9.69**. Press **O** to frame the object in the editor view. Press **NB** to enable the **Gouraud Shading (Lines)** mode, Fig. T3.

2. Press **C** to make **cabinetGeo** editable. Press **UB** to activate the **Ring Selection** 🖾 tool and then select the edge ring, refer Fig. T4. Press **MM** to connect the edges. Select the newly created edge loop by double-clicking on it using the **Move** tool and then slide it towards the negative X axis, Fig. T5.

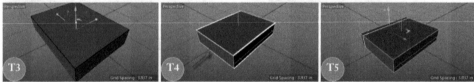

3. Select the edge ring using the **Ring Selection** tool, Fig. T6. Press **MM** thrice to create new edge loops, Fig. T7. Select the polygons that will make pages of the book, Fig. T8.

4. Press **MT** to enable the **Extrude** tool and then in the **Attribute Manager | Extrude | Options** tab, enter **-0.1** in the **Offset** field and **180** in the **Maximum Angle** field to extrude the polygons inwards, Fig. T9. In the **Front** view, select the points shown in Fig. T10. Make sure the **Only Select Visible Elements** check box is cleared in **Attribute Manager**. Move the points, as shown in Fig. T11.

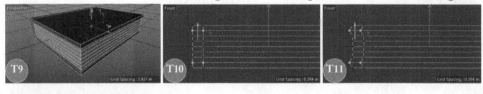

5. Similarly, move other points to make shape of the book, Fig. T12. Select the outer edges of the book, Fig. T13.

6. Press **MS** to activate the **Bevel** tool and then in **Attribute Manager | Bevel | Tool Option** tab, set **Offset** to **0.01** and **Subdivision** to 1 to bevel the edges, Fig. T14.

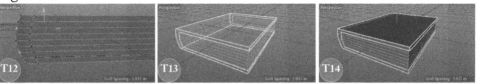

Tutorial 9: Creating a Waste Bin

In this tutorial, we will create model of a waste bin [Fig. T1].

The following table summarizes the tutorial:

Table T9	
Flow: The following sequence will be used to create the waste bin model: **1.** Create a **Cylinder** and then extrude polygons to create main shape. **2.** Use the **Cylinder** and **Torus** primitives to create the lid.	
Difficulty level	Beginner
Estimated time to complete	30 Minutes
Topics	• Getting Started • Creating the Bin
Resources folder	**chapter-m4**
Tutorial units	**Inches**
Final tutorial file	**m4-tut9-finish.c4d**

Getting Started

Start a new scene in CINEMA 4D and set units to **Inches**.

Creating the Bin

Follow the steps given next:

1. Choose **Cylinder** from the **Standard** palette | **Object** command group. In the **Attribute Manager | Cylinder | Object** tab, set **Radius** to 15, **Height** to 45,

Height Segments to **30**, and **Rotation Segments** to **50**. Press **NB** to enable **Gouroud Shading (Lines)** display mode. Press **C** to make **Cylinder** editable.

2. Switch to **Polygons** mode. Press **UB** to activate the **Ring Selection** tool and click at the top of the cylinder to select polygons. Press **Delete** to remove the top polygons, Fig. T2. Press **UL** to activate the **Loop Selection** tool and then select top and bottom rows of polygons using **Shift**, Fig. T3.

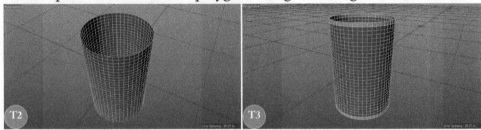

3. Press **MT** to enable the **Extrude** tool and then in the **Attribute Manager | Extrude | Options** tab, set **Offset** to **1.2**, Fig. T4. Ensure the newly extruded polygons are selected and then press **MS** to activate the **Bevel** tool. In **Attribute Manager | Bevel | Tool Option** tab, set **Offset** to **0.6**. In the **Attribute Manager | Bevel | Polygon Extrusion** tab, set **Extrusion** to **0.6** Fig. T5.

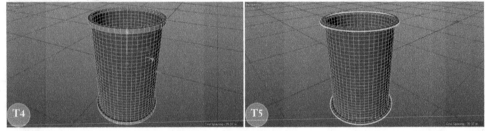

4. Press **UB** to activate the **Ring Selection** tool and then select every alternate column of polygons using **Shift**, Fig. T6. Activate the **Rectangle Selection** ☐ tool from the **Selection** command group and then in **Attribute Manager**, turn off the **Only Select Visible Elements** switch. In the **Front** view, remove two top and bottom loops of polygons using **Ctrl**, Fig. T7.

5. Press **MT** to enable the **Extrude** tool and then in **Attribute Manager | Extrude | Options** tab, set **Offset** to **-0.5**, Fig. T8. Press **NA** to enable the **Gouroud Shading** display mode. Hold **Alt** and then choose **Subdivision Surface** from the **Generators** command group to make the cylinder smooth, Fig. T9. In **Object Manager**, rename **Subdivision Surface** as **Waste Bin**.

What next?
Next, we will create a lid for the cylinder.

6. Choose **Cylinder** from the **Standard** palette | **Object** tab. In **Attribute Manager** | **Cylinder** | **Object** tab, set **Radius** to **17**, **Height** to **2**, **Height Segments** to **1**, and **Rotation Segments** to **50**. Press **NB** to enable **Gouroud Shading (Lines)** display mode. Rename **Cylinder** as **Lid** in **Object Manager** and then press **C** to make it editable.

7. Press **UB** to activate the **Ring Selection** tool and click at the bottom of the **Lid** to select bottom polygons. Press **MW** to select the **Extrude Inner** tool and then in the **Attribute Manager** | **Extrude Inner** | **Options** tab, set **Offset** to **1**, Fig. T10. Now, using **Extrude** tool extrude the polygons by setting **Offset** to **-1** in **Attribute Manager**, Fig. T11.

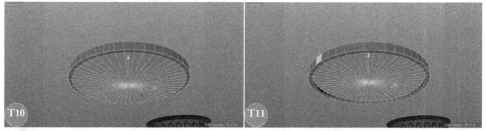

8. Select all points of the **Lid** in the **Points** mode and then press **UO** to optimize the **Lid**.

What just happened?
*I have welded the points of the cap of the **Lid** with rest of geometry using the **Optimize** function. Now, when I will apply the **Bevel** tool in up next, the whole geometry will remain intact.*

9. Select the edges loops shown in Fig. T12 and then press **MS** to activate the **Bevel** tool. In **Attribute Manager** | **Bevel** | **Tool Option** tab, set **Offset** to **0.2** and **Subdivision** to **2**, Fig. T13.

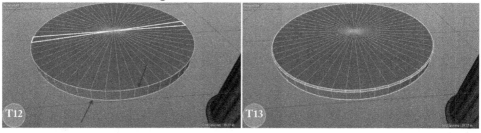

10. Using the **Ring Selection** tool, select the top polygons of the **Lid** and then press **MW** to select the **Extrude Inner** tool and then **in Attribute Manager | Extrude Inner | Options** tab, set **Offset** to **9.5**, Fig. T14.

11. Activate the **Move** tool and then move the selected polygon in the positive Y direction by **1** unit, Fig. T15. Press **NA** to enable the **Gouroud Shading** display mode and then align the **Lid** on top of the **Waste Bin**, Fig. T16. Now, we will create handle for the **Lid**.

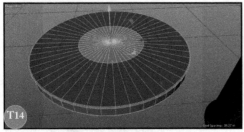
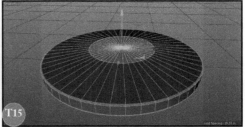

12. Choose **Torus** ⊚ from the **Standard** palette | **Object** command group. In the **Attribute Manager | Torus | Object** tab, set **Ring Radius** to **4.64**, **Ring Segments** to **50**, **Pipe Radius** to **0.84**, and **Pipe Segments** to **36** and then align it with the **Lid**, Fig. T17. Rename **Torus** as **Handle** in **Object Manager**. You can also use the options available in the **Slice** tab to create half torus and then align it with the lid.

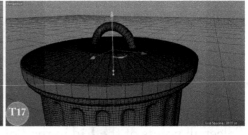

Tutorial 10: Creating a Desk

In this tutorial, we will create model of a waste bin [Fig. T1].

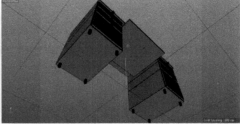

The following table summarizes the tutorial:

Table T10
Flow: The following sequence will be used to create the desk:
1. Use the **Cube** primitives to create the basic shape of the desk **2.** Edit elements to create drawers. **3.** Use the **Cylinder** primitive to create knobs for the drawers.

Difficulty level	Intermediate

Table T10	
Estimated time to complete	30 Minutes
Topics	• Getting Started • Creating the Desk
Resources folder	**chapter-m4**
Tutorial units	**Inches**
Final tutorial file	**m4-tut10-finish.c4d**

Getting Started

Start a new scene in CINEMA 4D and set units to **Centimeters**.

Creating the Desk

Follow the steps given next:

1. Choose **Cube** from the **Standard Palette | Object** command group to create a cube in the editor view. In the **Attribute Manager | Cube | Object** tab, set **Size X** to **60**, **Size Y** to **2.5**, and **Size Z** to **150**. Press **O** to frame the object in the editor view. Press **NB** to enable the **Gouraud Shading (Lines)** mode. Create another cube and then in the **Attribute Manager | Cube.1 | Object** tab, set **Size X** to **60**, **Size Y** to **62**, and **Size Z** to **40**, Fig. T2.

2. Select **Cube** and **Cube.1** in **Object Manager** and then choose **Arrange Objects | Center** from the **Tools** menu. In the **Attribute Manager | Center | Options** tab, select **Negative** from the **Y Axis** and **Z Axis** drop-downs. Click **Apply** from the **Tool** tab. Click **New Transform** from the **Tool** tab and then in the **Options** tab, select **Positive** from the **Y Axis** drop-down to align the cubes, Fig. T3. Now, using the **Move** tool align the two cubes as shown in Fig. T4.

3. Ensure **Cube.1** selected in **Object Manager** and then choose **Arrange Objects | Duplicate** from the **Tools** menu. In the **Attribute Manager | Duplicate | Duplicate** tab, set **Copies** to **1**. In the **Options** tab, select **Linear** from the **Mode** drop-down. In the **Position** section, set **Move X**, **Move Y**, and **Move Z** to **0**, **0**, and **110**, respectively to align the duplicate to the other end of the table top, Fig. T5.

What just happened?
*Here, I've created a duplicate of the **Cube.1** using the **Duplicate** command and then offset it by **110** [150-40=110] units along the positive Z axis.*

4. Press **Ctrl+A** to select all cubes and then press **C** to make them editable. Now, choose **Conversion | Connect Objects + Delete** from the **Mesh** menu. Rename the unified geometry as **deskGeo** in **Object Manager**.

What just happened?
I've made all thee cube primitives editable and then combined the result in a single unified polygon object.

5. Press **UB** to activate the **Ring Selection** tool and then select the edges [in the **Edges** mode] of the bases of the table using **Shift**, Fig. T6. Press **MM** to connect the edges, Fig. T7. Now, select the top rings of the bases, Fig. T8, and then press **MM** to connect the edges, Fig. T9.

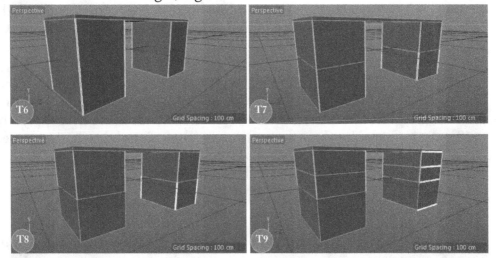

6. Select the polygons shown in Fig. T10. Press **MW** to select the **Extrude Inner** tool and then in the **Attribute Manager | Extrude Inner | Options** tab, clear the **Preserve Groups** check box and then set **Offset** to **1.5**, Fig. T11. Press **MT** to enable the **Extrude** tool and then in the **Attribute Manager | Extrude | Options** tab, set **Offset** to **1.5**, Fig. T12.

What next?
*Now, we'll create the keyboard support. To do so, we will place some edges using the **Plane Cut** tool.*

7. Activate the **Right** viewport and then press **MJ** to activate the **Plane Cut** tool. Now, click and drag to create two edge loops using **Shift**, Fig. T13.

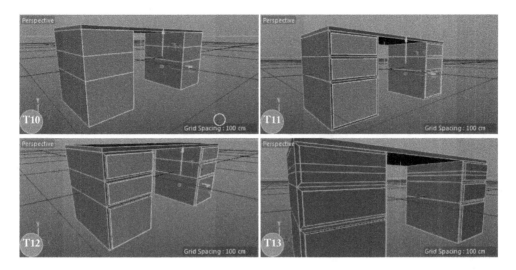

→ **What next?**
*Now, select the polygons shown in Fig. T14 and press **UP** to split the polygons.*

8. Rename the new geometry is **keyboardGeo** in **Object Manager**. Select the newly created polygons and press **MB** to activate the **Bridge** tool. In the **Attribute Manager | Bridge | Options** section, clear the **Delete Original Polygons** check box and then click and drag on one of the polygons to make a bridge between them, Fig. T15. Now, switch to the **Model** mode and pull out the keyboard little bit using the **Move** tool, Fig. T16.

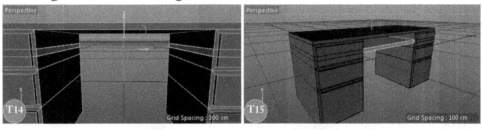

9. Press **Ctrl+A** to select all objects in **Object Manager** and then choose **Conversion | Connect Objects + Delete** from the **Mesh** menu. Rename the unified geometry as **deskGeo** in **Object Manager**.

✎ *Note: Cleaning the Model*
*You can remove the edges that we added for creating the keyboard support by first selecting them and then using the **Dissolve** command.*

10. Select the polygon shown in Fig. T17 and then press **MT** to enable the **Extrude** tool and then in the **Attribute Manager | Extrude | Options** tab, set **Offset** to **2.5**, Fig. T18. In the **Edges** mode, press **Ctrl+A** to select all edges and then press **MS** to activate the **Bevel** tool. In the **Attribute Manager | Bevel | Tool Option** tab, set **Offset** to **0.1** and **Subdivision** to **2**, Fig. T19.

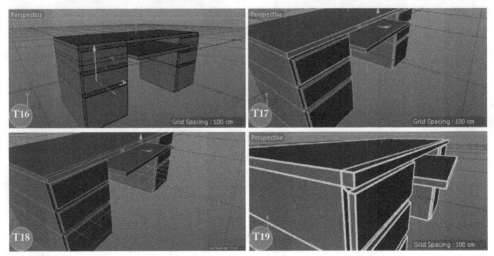

→ **What next?**
Now, we'll create knobs for the drawers using a cylinder.

11. Choose **Cylinder** from the **Standard** palette | **Object** tab. In the **Attribute Manager** | **Cylinder** | **Object** tab, set **Radius** to **1.5**, **Height** to **6**, and **Rotation Segments** to **18**. In the **Caps** tab, select the **Fillet** check box and then set **Segments** to **3**, **Height Segments** to **1**, and **Radius** to **0.074**. Press **C** to make the cylinder editable. Select the edge ring, shown in Fig. T20 and then press **MM** to connect the edges, Fig. T21.

12. Select the newly created edge and then press **MS** to activate the **Bevel** tool. In the **Attribute Manager** | **Bevel** | **Tool Option** tab, set **Offset** to **0.05** and **Subdivisions** to **0**, Fig. T22. **Ctrl+click** on **Polygons** in **Tools** palette to select the newly created polygons. Now, interactively inset and extrude the polygons using the **Extrude Inner** and **Extrude** tools, respectively, Fig. T23. Make sure the **Preserve Group** check box is selected when you use the **Extrude Inner** and **Extrude** tools.

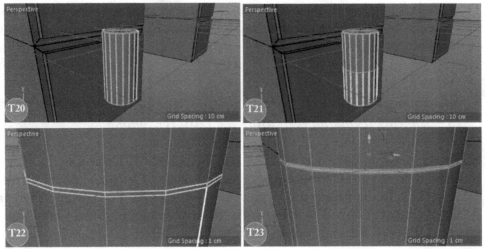

13. Ensure **Cylinder** is selected in **Object Manager** and then choose **Subdivision Surface** from the **Generator** command group with **Alt** held down to smooth the cylinder. Now, create copies of the cylinder and align them with drawers of the desk, Fig. T24. Similarly, create legs of the desk, Fig. T25.

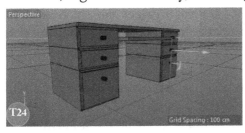 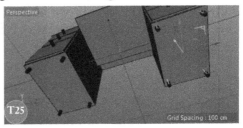

Tutorial 11: Creating an Exterior Scene

In this tutorial, we will model an exterior scene using various modeling techniques [Fig. T1].

The following table summarizes the tutorial:

Table T11	
Flow: The following sequence will be used to create the exterior scene: **1.** Use **Cube** primitives and polygon editing tools to create the basic shape of the building. **2.** Use the **Circle** primitives and the **Loft** generator to create light poles. **3.** Use the **Cube** and **Cylinder** primitives to create the light sockets. **4.** Use the **Cube** primitives to create the doors and **Torus** primitives to create the door handles.	
Difficulty level	Beginner
Estimated time to complete	90 Minutes
Topics	• Getting Started • Creating the Scene
Resources folder	**chapter-m4**
Tutorial units	**Meters**
Final tutorial file	**m4-tut11-finish.c4d**

Getting Started

Start a new scene in CINEMA 4D and set units to **Meters**.

Creating the Scene

Follow the steps given next:

1. Choose **Cube** from the **Standard** palette | **Object tab** to create a cube in the editor view. In the **Attribute Manager** | **Cube** | **Object** tab, set **Size X** to **20**, **Size Y** to **8**, and **Size Z** to **60**. Press **NB** to enable the **Gouraud Shading (Lines)** mode.

2. Create another **Cube** object and then in the **Attribute Manager** | **Cube** | **Object** tab, set **Size X** to **14**, **Size Y** to **8**, and **Size Z** to **52**. Also, set **Segments X**, **Segments Y**, and **Segments Z** to **4**, **1**, and **14**, respectively. Now, align the **Cube.1** to the bottom of **Cube** [Fig. T2]. Choose **Filter** | **Grid** from the MEV menu to turn off the grid. Select **Cube** in **Object Manager** and then press **C** to make it editable. Similarly, make **Cube.1** editable. Select **Cube** in **Object Manager** and activate the **Edges** mode from the **Tools** palette.

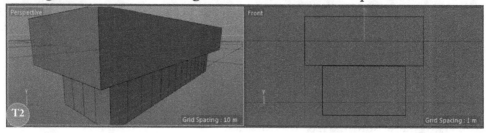

3. Press **ML** to enable the **Loop/Path Cut** tool. Create an edge loop. In the **Attribute Manager** | **Options** tab, set Offset to 25%, , Fig. T3. Now, select the front polygons, Fig. T4, and then press **MW** to select the **Extrude Inner** tool and then in the **Attribute Manager** | **Extrude Inner** | **Options** tab, clear the **Preserve Groups** check box and then set **Offset** to **0.6**, Fig. T4.

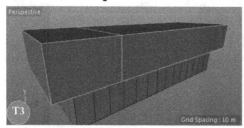
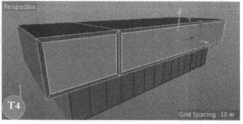

4. Press **MT** to enable the **Extrude** tool and then in the **Attribute Manager** | **Cube**| **Extrude** | **Options** tab, set **Offset** to -5, Fig. T5. Choose **Cube** from the **Standard** palette | **Object** command group to create a cube in the editor view. In the **Attribute Manager** | **Cube** | **Object** tab, set **Size X** to **4.61**, **Size Y** to **6.827**, and **Size Z** to **0.602**. Align the cube, Fig. T6.

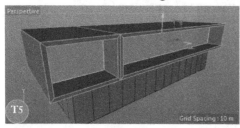
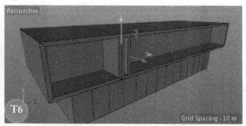

5. Make sure **Cube.2** is selected in **Object Manager** and then choose **Arrange Objects | Duplicate** from the **Tools** menu. In the **Attribute Manager | Duplicate | Duplicate** tab, set **Copies** to **39**, and **Clone Mode** to **Instances**. From the **Options** tab, set **Mode** to **Linear** and then in the **Options** tab | **Position** section, set **Move XYZ** to **0, 0**, and **1.1**, respectively, Fig. T7. Select **Cube.1** in **Object Manager** and then select the polygon shown in Fig. T8.

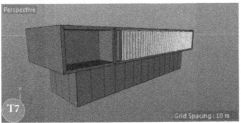 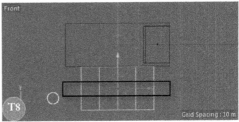

6. Press **MW** to select the **Extrude Inner** tool and then in the **Attribute Manager | Extrude Inner | Options** tab, clear the **Preserve Groups** check box and then set **Offset** to **0.22**. Press **MT** to enable the **Extrude** tool and then in the **Attribute Manager | Extrude | Options** tab, set **Offset** to **-0.25**, Fig. T9. Ensure polygons are still selected and then choose **Set Selection** from the **Select** menu to create a selection set. In the **Attribute Manager | Basic Properties** section, **glassSelection** in the **Name** field to name the selection.

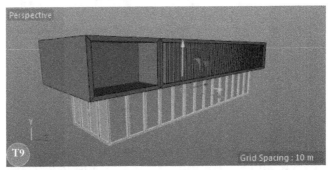

What just happened?

*Here, I've frozen the polygon selection. You can also create selection set for the point and edge selections. When you create a selection set, you can recall that selection later. For example, while texturing, if you want to apply glass material to these polygons, you can easily recall the selection by double-clicking on the selection tag in **Object Manager**. You can freeze more than 10 selections per object, however, many of the commands operate on the first 10 sets only.*

7. Choose **Plane** from the **Standard** palette | **Object** command group to create a plane in the editor view that will act as ground. In the **Attribute Manager | Plane | Object** tab, set **Width** to **600** and **Height** to **800**, and then align it as shown in Fig. T10. Choose **Cube** from **Standard** palette | **Object** command group to create a cube in the editor view. In the **Attribute Manager | Cube.3 | Object** tab, set **Size X** to **36, Size Y** to **0.2**, and **Size Z** to **71**. Now, align it with plane, Fig. T11.

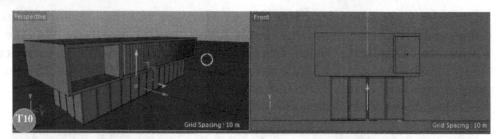

8. Choose **Rectangle** from the **Standard** palette | **Spline** command group to create a rectangle in the editor view. In the **Attribute Manager** | **Rectangle** | **Object** tab, set **Width** and **Height** to **12** and **48**, respectively. Select the **Rounding** check box and then set **Radius** to **4** and **Plane** to **XZ**. Ensure **Rectangle** is selected in **Object Manager** and then choose **Extrude** from the **Standard** palette | **Generators** command group with **Alt** held down. In the **Attribute Manager** | **Extrude** | **Object** tab, set **Movement XYZ** to **0, 0.4,** and **0**, respectively. Now, align object with the ground plane, Fig. T12.

(→) *What next?*
Now, we will create the light pole.

9. Choose **Circle** from the **Standard** palette | **Spline** command group to create a circle in the editor view. In the **Attribute Manager** | **Object** tab, set **Radius** to **0.5** and **Plane** to **XZ**. Now, align it, as shown in Fig. T13. Create three more copies of the **Circle** and set **Radius** to **0.4, 0.3,** and **0.2**, respectively. Align the circles, as shown in Fig. T14.

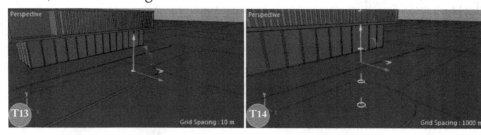

10. Choose **Loft** from the **Generators** command group. In **Object Manager**, drag all circle objects onto it to make them children of the **Loft** object, the pole geometry is created in the viewport, Fig. T15. Choose **Cylinder** from the **Standard** palette | **Object** command group. In the **Attribute Manager** | **Cylinder** | **Object** tab, set **Radius** to **0.1**, **Height** to **3.168**, **Height Segments** to **1**, and then align it with the pole, Fig. T16.

→ **What next?**
Now, we will create lights.

11. Choose **Cube** from the **Standard** palette | **Object** command group to create a cube in the editor view. In the **Attribute Manager** | **Cube.4** | **Object** tab, set **Size X** to **1.39**, **Size Y** to **0.71**, and **Size Z** to **0.931**. Press **C** to make **Cube.4** editable. Select the points, as shown in Fig. T17 and then move them down, Fig. T18.

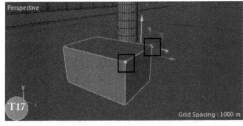
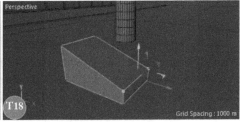

12. Select all polygons of the **Cube.4** and then press **MW** to select the **Extrude Inner** tool and then in the **Attribute Manager** | **Extrude Inner** | **Options** tab, clear the **Preserve Groups** check box and then set **Offset** to **0.04**. Press **UI** to invert the selection and then press **MT** to enable the **Extrude** tool. In the **Attribute Manager** | **Extrude** | **Options** tab, select the **Preserve Groups** check box. Set **Offset** to **0.018** and **Maximum Angle** to **180**, Fig. T19.

13. Ensure polygons are still selected and then choose **Set Selection** from the **Select** menu to create a selection set. In the **Attribute Manager** | **Basic Properties** section, type **lightCover** in the **Name** field to name the selection. Align **Cube.4** with the pole, refer Fig. T20. Create a copy of **Cube.4** and align it with the other side of the pole, Fig. T20.

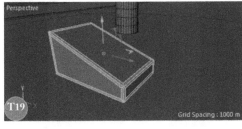
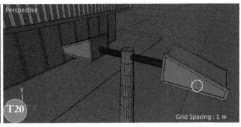

14. Select the **Loft, Cylinder, Cube.4,** and **Cube.5** objects in **Object Manager** and press **Alt+G** to group the objects. Rename **Null** as **Pole.1** in **Object Manager**. Make a duplicate of **Pole.1** and then rename it as **Pole.2**. Align **Pole.2**, as shown in Fig. T21.

15. Create main door of the building using **Box** and **Torus** primitives, Fig. T22.

→ *What next?*
   Now, let's create the logo of the company.

16. Choose **Text** from the **Standard** palette | **Spline** command group to create a **Text** object in the editor view. In the **Attribute Manager** | **Text** | **Object** tab, set **Text** to **My Inc.** Choose a font of your choice from the **Font** drop-down. Set **Height** to **3**. You might have to adjust the height as per the font you have chosen. Align text, as shown in Fig. T23.

17. Ensure **Text** is selected in **Object Manager** and then choose **Extrude** from the **Standard** palette | **Generators** command group with **Alt** held down. In the **Attribute Manager** | **Extrude** | **Object** tab, set **Movement XYZ** to **0, 0**, and **0.3**, respectively. In the **Caps** tab, select **Fillet Cap** from the **Start** and **End** drop-downs. Set **Start Steps/Radius** to **3** and **0.05** and **End Steps/Radius** to **2** and **0.02**. Set **Fillet Type** to **Engraved**, Fig. T24.

Tutorial 12: Creating a Volleyball Model

In this tutorial, we will create a volleyball model using the polygon modeling techniques [Fig. T1].

The following table summarizes the tutorial:

Table T12	
Flow: The following sequence will be used to create the volleyball model: **1.** Use the **Hexahedron** sphere type to create the basic shape and **2.** Use the polygon editing tools to create the final shape.	
Difficulty level	Intermediate
Estimated time to complete	30 Minutes
Topics	• Getting Started • Creating the Volleyball Model
Resources folder	**chapter-m4**
Tutorial units	**Centimeters**
Final tutorial file	**m4-tut12-finish.c4d**

Getting Started
Start a new scene in CINEMA 4D and set units to **Centimeters**.

Creating the Volleyball Model
Follow the steps given next:

1. Choose **Sphere** from the **Standard** palette | **Object** command group to create a sphere in the editor view. Press **NB** to enable the **Gouraud Shading (Lines)** mode. On the **Attribute Manager | Sphere | Object** tab, set **Segments** to **27** and **Type** to **Hexahedron**. Press **C** to make object editable.

2. Activate the **Edges** mode and then press **UL** to invoke the **Loop Selection** tool. Now, select the loop, as shown in Fig. T2. Press **UF** to invoke the **Fill Selection** tool and then select the polygons [Fig. T3]. Now, press **Delete** to remove the selected polygons [Fig. T4].

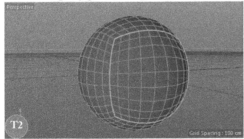 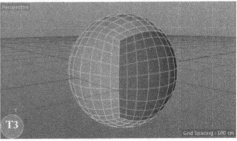

> *What just happened?*
> *Here, I have first created a closed edge boundary selection using the **Loop Selection** tool and then applied the **Fill Selection** tool to the other side of the boundary. This tool allows you to create a polygon selection from an existing edge selection [**preferably closed**].*

3. Activate the **Model** mode and then choose **Mesh | Command | Optimize** from the menubar.

> **?**
>
> *What just happened?*
> *When I selected the polygons and deleted them, CINEMA 4D did not delete the points from the 3D space [Fig. T5]. The **Optimize** command allows to remove such left-over points and gives you clean geometry. You can also optimize edges and polygons as well. This command also works on the spline points.*

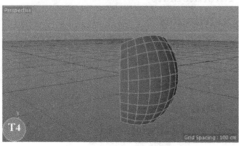 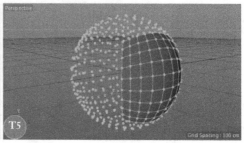

4. Activate the **Polygons** mode and then make a selection [Fig. T6]. Press **UD** to execute the **Disconnect** command.

> **?**
>
> *Why did you disconnect polygons?*
> *If you observe a real-world volleyball, you would notice that the volleyball is made up of 18 different patches. Disconnecting the patches will help us in selecting them easily when we will apply textures to them.*

5. Press **MT** to invoke the **Extrude** tool. Now, on the **Attribute Manager | Extrude | Tool** tab, click **Apply** to apply a **5** units extrusion to the selected polygons [Fig. T7].

6. Make a polygon selection, as shown in Fig. T8 and then extrude by **5** units [Fig. T9]. Press **Ctrl+A** to select all polygons and then press **MY** to invoke the **Smooth Shift** tool. On the **Attribute Manager | Smooth Shift | Tool** tab, click **Apply**. Set **Maximum Angle** to **25** and **Offset** to **0.1**.

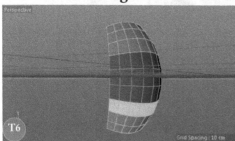 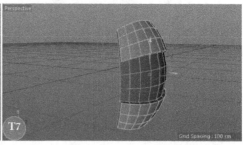

> **?**
>
> *What just happened?*
> *Here, I have used the **Smooth Shift** tool which is similar to the **Extrude** tool. However, when you use **Smooth Shift**, the selected surfaces will be moved in the direction of the normals. The value of the **Maximum Angle** parameter controls if a new connection surface would be created between the polygons.*

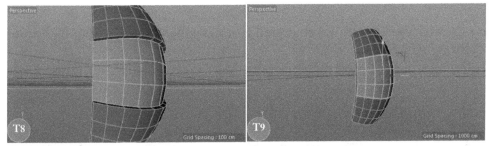

7. Activate the **Model** mode and ensure that **Sphere** is selected in **Object Manager**. Now, choose **Subdivision Surface** from the **Standard** palette | **Generators** command group with **Alt** down to make sphere child of the **Subdivision Surface** object [Fig. T10].

8. Ensure that the **Subdivision Surface** object is selected in **Object Manager** and then choose **Mesh | Conversion | Current State to Object** from the menubar to bake the subdivisions into the sphere. Now, delete the **Subdivision Surface** object from **Object Manager**. Press **NA** to enable the **Gouraud Shading** mode. Choose **Enable Quantizing** from the **Snap** menu or press **Shift+Q**.

> **What just happened?**
> The **Enable Quantizing** option allows you to restrict the stepless motion to a defined grid. We will be creating copies of the existing geometry and then aligning them using the **Rotate** tool. The quantizing settings will allow us to rotate the geometry in fix increments. The default value for the rotation is **10** degrees. You can change this value from the **Modeling** mode page in **Attribute Manager**.

9. Active the **Rotate** tool and then create and align copies using **Ctrl** [Fig. T11]. Now, select all objects in **Object Manager** and then choose **Mesh | Conversion | Connect Objects + Delete** to create a single object. Rename the object as **Volleyball**.

Tutorial 13: Creating a Shattered Abstract Sphere

In this tutorial, we will create a shattered abstract sphere model using the **Explosion FX** deformer [Fig. T1].

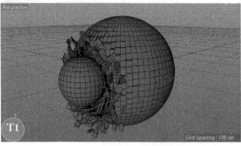

The following table summarizes the tutorial:

Table T13	
Flow: The following sequence will be used to create the abstract sphere: **1.** Create a **Sphere** primitive, **2.** Use the **Explosion FX** deformer to shatter the sphere. **3.** Create a smaller sphere and place it near the shattered area.	
Difficulty level	Intermediate
Estimated time to complete	30 Minutes
Topics	• Getting Started • Creating the Sphere
Resources folder	**chapter-m4**
Tutorial units	**Centimeters**
Final tutorial file	**m4-tut13-finish.c4d**

Getting Started

Start a new scene in CINEMA 4D and set units to **Centimeters**.

Creating the Sphere

Follow the steps given next:

1. Choose **Sphere** from the **Standard** palette | **Object** command group to create a sphere in the editor view. Press **NB** to enable the **Gouraud Shading (Lines)** mode. On the **Attribute Manager | Sphere | Object** tab, set **Segments** to **64**. Create another sphere and then on the **Attribute Manager | Sphere.1 | Object** tab, set **Radius** to **50** and **Segments** to **32**. Place the spheres, as shown in Fig. T2.

2. Select **Sphere** in **Object Manager** and then choose **Explosion FX** from the **Standard** palette | **Deformer** command group with **Shift** held down to make the **Explosion FX** object child of **Sphere**. Notice in the viewport the **Explosion FX** gizmo appears in the view and sphere is shattered into pieces [refer to Fig. T3].

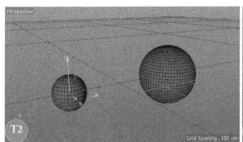 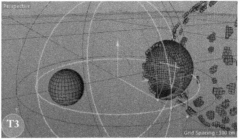

? *What are these elliptical concentric rings on gizmo control?*

*These concentric rings are controls that you can use to define various parameters interactively in the editor view. The innermost green rings control the **Time** parameters found in the **Attribute Manager | Object** tab. You can use this parameter to animate the explosion. The red rings control the **Blast Range** parameter which is found in the **Attribute Manager | Explosion** tab. The objects that are outside the red rings [blast range] are not accelerated by the blast. The blue rings control the **Range** parameter which is found in the **Attribute Manager | Gravity** tab. These rings define an area in which the objects will be affected by the gravity. However, all objects inside the blast range defined by red ring will be affected by the gravity regardless.*

3. On the **Attribute Manager | Coord** tab, set **P.Z** to **-245**. On the **Attribute Manager | Object** tab, make sure **Time** is set to **10%**. On the **Explosion** tab, set **Strength** to **40, Decay** to **70, Variation** to **55, Direction** to **Except Z, Blast Speed** to **45, Decay** to **9,** and **Blast Range** to **180**.

? *What just happened?*

*Here, I've set the explosion parameters. The **Time** parameter can be used to animate the explosion. **Strength** controls the force used to accelerate the clusters. **Decay** sets a falloff for the Strength. Variation controls vary the strength for each cluster. Here, **Direction** is set to **Except Z**. As a result, the acceleration of the clusters will be in all directions except the **Z** axis. The **Blast Speed** parameter controls the blast speed. A cluster remain in place until the blast reaches to it. The green rings in the editor view represent the blast. If you set this parameter to **0**, all clusters will be accelerated immediately. The blast speed is measured in meters per second. Decay is the falloff for the **Blast Speed**.*

4. On the **Cluster** tab, set **Thickness** to **2, Min Polys** to **1,** and **Max Polys** to **2**. On the **Rotation** tab, set **Speed** to **100, Decay** to **30,** and **Variation** to **10**. Also, set **Rotation Axis Variation** to **10** [see the effect of these values in Fig. T4]. Move **Sphere.1,** as shown in Fig. T5.

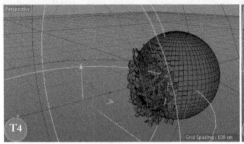

What just happened?
Here, I've set the parameters for the clusters. **Thickness** extrudes the clusters in the direction of normals to give them thickness. If you want to reverse the direction of the extrude, enter a negative value for this parameter. The **Min Polys** and **Max Polys** parameters define how cluster will be formed out of the object's polygons. If you want to create a cluster per polygon, choose **Polygons** from the **Cluster Type** drop-down. After setting the parameters for the clusters. I've changed values in the **Rotation** tab to give random rotation to the clusters.

Now, you can texture the scene.

Quiz

Evaluate your skills to see how many questions you can answer correctly.

Multiple Choice
Answer the following questions, only one choice is correct.

1. Which of the following key combinations is used to invert the current selection?

   [A] UI   [B] UX
   [C] UJ   [D] UK

2. Which of the following key combinations is used to select all points, edges, or polygons connected to the selected element?

   [A] UX   [B] UW
   [C] UJ   [D] UK

3. Which of the following tools allows you to create connections between the unconnected surfaces?

   [A] Connect   [B] Bridge
   [C] Loft   [D] Sweep

4. Which of the following keys is associated with the **Extrude** tool?

[A] D   [B] E
[C] I   [D] H

Fill in the Blanks
Fill in the blanks in each of the following statements:

1. CINEMA 4D provides three modes for polygonal modeling: _____, _____, and _____.

2. To make a polygon object editable, press _____.

3. You can use the _____ command to store selection sets and then recall them later.

True or False
State whether each of the following is true or false:

1. The tools and commands available in the **Select** menu are also available in the U hidden menu.

2. The **Path Selection** tool works only in the **Polygons** mode.

Practice Activities

Activity 1: Create a Solid Model

Create the model shown in Fig. A1. Rest of the figures show hints for creating the model.

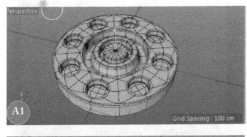

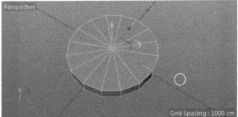

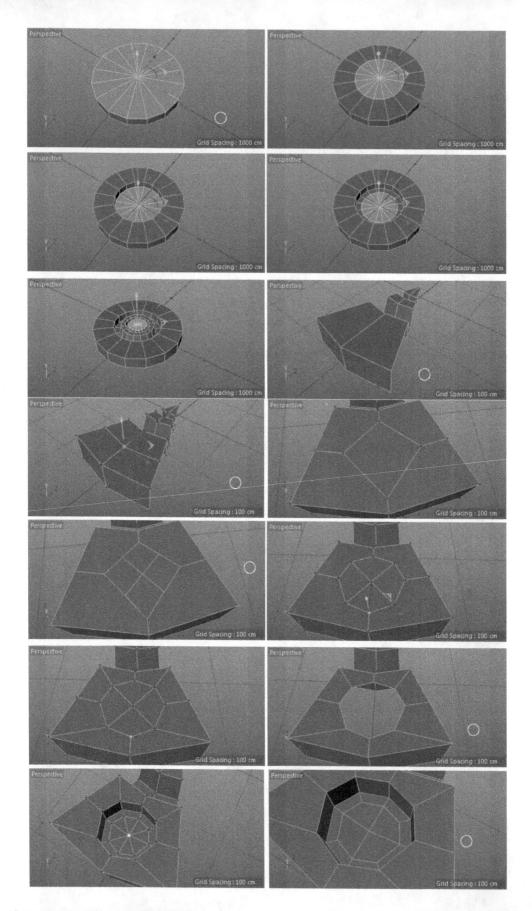

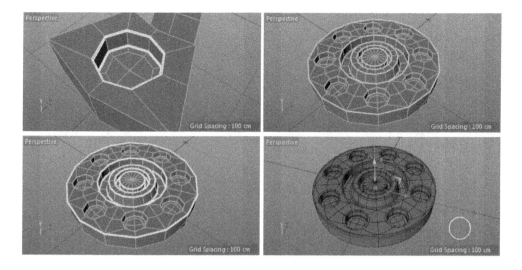

Activity 2: Creating a Flash Drive

Create a model of flash drive using polygon modeling techniques [see Fig. A2].

Hint Activity - 2

*Set **Units** to **Millimeters**. Create a **Cylinder** primitive and then set its **Radius** to 7.5, **Height** to 7, **Height Segments** to 1, and **Rotation Segments** to 36. Make it editable and then dissolve the top and bottom points using the **Dissolve** command, see Fig. A3. Weld the points using the **Connect** object, as done in **Tutorial -1**. Select the one half of the points in the **Top** view and then move them about 25 units to the right see Fig. A4. Now, use various modeling tools and functions to create the flash drive model.*

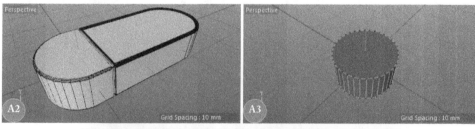

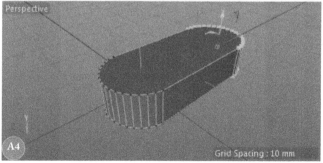

Activity 3: Creating a USB Connector

Create a model of the USB connector using polygon modeling techniques [see Fig. A5].

Hint Activity - 3

Set **Units** to **Millimeters**. Create a **Box** and then set its **Size X**, **Size Y**, and **Size Z** to **15**, **5**, and **30**, respectively. Now, use various modeling tools and functions to create the USB connector model.

Activity 4: Creating a Kitchen Cabinet

Create the kitchen cabinet model [see Fig. A6] using the **Cube** primitive. Use dimensions of your choice.

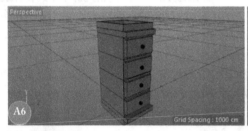

Summary

This chapter covered the following topics:

- Polygons components
- Polygon modeling techniques
- Selection tools
- Polygons structure tools
- Modeling Objects
- Deformers

Chapter MBT: Bonus Tutorials [Modeling]

Tutorials

Before you start the tutorials, create a folder with the name **chapter-mbt**. We'll use this folder to host all the tutorial files and other resources.

Tutorial 1: Creating a Chair

In this tutorials, we will create model of a chair, as shown in Fig. T1.

The following table summarizes the tutorial:

Table T1	
Flow: The following sequence will be used to create the chair model: **1.** Create left profile and right profile curves of the tool using the **Pen** tool. **2.** Create the right profile by making a copy of the left profile. **3.** Create frame using the **Circle** spline and **Sweep** generator. **4.** Use a **Cube** primitive to create seat. **5.** Use a **FFD** deformer to bend the seat.	
Difficulty level	Beginner
Estimated time to complete	30 Minutes
Topics	• Getting Started • Creating the Chair
Resources folder	**chapter-mbt**
Tutorial units	**Inches**
Final tutorial file	**mbt-tut1-finish.c4d**

Getting Started
Start a new scene in CINEMA 4D and set units to **Inches**.

Creating the Chair
Follow the steps given next:

1. Choose **Pen** from the **Standard** palette | **Spline** command group to create a shape in the **Front** view [Fig. T2]. Select the point, as shown in Fig. T3.

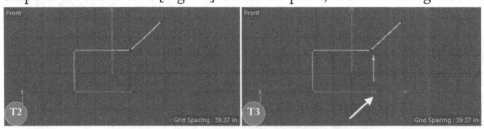

2. On **Coordinate Manager**, enter **9.886, 0.086**, and **-10.75** in the X, Y, and Z fields of the **Position** parameter, respectively. Similarly, set the other points using the values shown in Table T2.1.

Table T1.1 - Coordinates for creating points			
Point	X	Y	Z
Ist	9.886	0.086	-10.75
2nd	-9.886	0.086	-10.75
3rd	-9.886	14.834	-10.75
4th	6.285	14.834	-10.75
5th	9.886	27.011	-10.75

After entering the values, the spline is shown in Fig. T4.

3. Choose **Circle** from the **Standard** palette | **Spline** command group to create a circle in the editor view. In the **Attribute Manager** | **Circle** | **Object** tab, set **Radius** to **0.4**. Make sure **Spline** is selected in **Object Manager** and then create a copy of it by **Ctrl** dragging it in the editor view about **21** units along the **Z** axis [Fig. T5]. Choose **Pen** from the **Standard** palette | **Spline** command group and then connect the two splines [Fig. T6]. Select the points shown in Fig. T7.

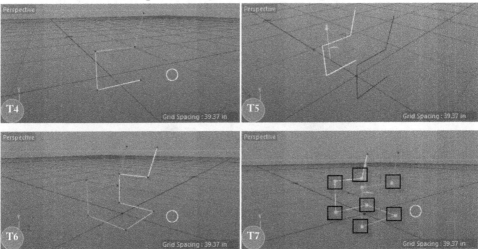

4. RMB click and choose **Chamfer** from the popup menu. Chamfer the points [Fig. T8]. Now, using the **Circle** and **Sweep** generators, create the frame of the chair [Fig. T9]. Rename **Sweep** as **Frame** in **Object Manager**.

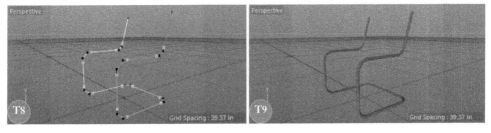

5. Create caps for frame using the **Tube** primitive [Fig. T10]. Select all objects in **Object Manager** and then press **Alt+G** to group them. Rename the group as **frameGrp**. Now, we'll create seat for the chair.

6. Choose **Cube** from the **Standard** palette | **Object** command group to create a cube in the editor view. In the **Attribute Manager | Cube | Object** tab, set **Size X** to **14.152**, **Size Y** to **2.5 in**, and **Size Z** to **24.283**, respectively. Also, set the **Segment** X, **Segment** Y, and **Segment** Z to **4, 3**, and **4**, respectively. Align the **Cube** with the frame [Fig. T11].

7. Add a **Subdivision Surface** generator to the **Cube** to smooth it. Now, press **C** to make the **Subdivision Surface** generator editable. Rename the object as **Seat** in **Object Manager**. Now, we'll create piping on the seat. Using the **Loop Selection** tool, select two edges as shown in Fig. T12.

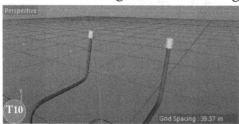
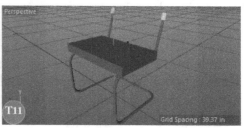

8. Choose **Commands | Edge to Spline** from the **Mesh** menu to create spline from the selected edges. Select **Seat.Spline** in **Object Manager** and drag it out of the **Seat** group to the top level [Fig. T13].

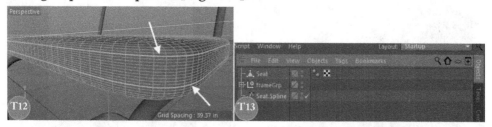

9. Create a copy of the **Circle** that you we created earlier and then set its **Radius** to **0.1**. Create piping geometry by using the **Sweep** generator [Fig. T14]. Use the commands available in the **Mesh | Conversion** menu to connect the elements and to make a single editable object for the seat and piping. Rename the object as **Seat** in **Object Manager** [Fig. T15]. Make sure **seat** is selected in **Object Manager** and then center the axis on the seat using the **L** key, in not already at the center.

10. Add a **FFD** deformer to the seat. Select the middle points of **FFD** and then move them along the negative **Y** axis to create a bend in the seat [Fig. T16]. Convert seat to a single object using the **Current State to Object** command. Delete the **Seat** object connected to the **FFD** deformer.

11. To make the back support of the chair, create a copy of the seat and then align it with the frame. If required, scale down the height of the back support [Fig. T17]. Select everything in **Object Manager** and then group the object as **Chair**.

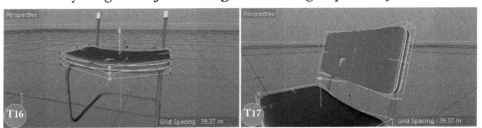

Tutorial 2: Creating a Chair

In this tutorial, we will create a chair using the spline and polygon modeling techniques [Fig. T1].

The following table summarizes the tutorial:

Table T2: Creating the Chair	
Flow: The following sequence will be used to create the chair model: 1. Create a **Cube** as reference for the chair. 2. Use the **Pen** tool, the **Circle** spline, and the **Sweep** deformer to create the frame. 3. Create and modify the **Plane** primitive to create the seat using the **Bevel** tool. 4. Use a **FFD** deformer to bend the shape of the seat.	
Difficulty level	Beginner
Estimated time to complete	45 Minutes
Topics	• Getting Started • Creating the Chair
Resources folder	**chapter-m1**
Tutorial units	**Inches**
Final tutorial file	**mbt-tut2-finish.c4d**

Getting Started
Start a new scene in CINEMA 4D and set units to **Inches**.

Creating the Chair
We'll first create a cube that will work like a template that will help us in the modeling process. Follow the steps given next:

1. Choose **Cube** from the **Standard** palette | **Object** command group to create a cube in the editor view. In the **Attribute Manager | Cube | Object** tab, set **Size X** to **20**, **Size Y** to **30**, and **Size Z** to **20**.

2. Set **Segment X**, **Segment Y**, and **Segment Z** to **1**, **2**, and **4**, respectively. Press **O** to frame the object in the editor view. Press **NB** to enable the **Gouraud Shading (Lines)** mode, Fig. T2.

3. Press **Shift+S** to enable snapping and then choose **Pen** from the **Standard** palette. Now, create a rectangular spline by clicking on the four corners of the lower half of the cube [Fig. T3].

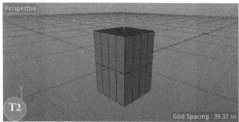
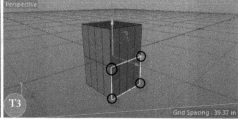

4. Move the spline slightly to the right of the cube so that it is visible clearly. Select the two top points and then chamfer them using the **Chamfer** tool [Fig. T4]. Also, chamfer the bottom points [Fig. T5].

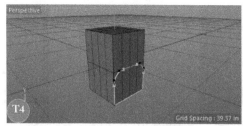
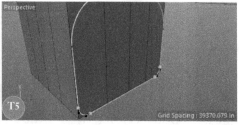

5. Create a circle of radius **0.3** and then create the frame of the chair using the **Sweep** generator [Fig. T6]. Create a cube and then set its **Size X** to **2.6**, **Size Y** to **1**, and **Size Z** to **1.5**. Also, select the **Fillet** check box and then set **Fillet Radius** to **0.3**. Align, it with the bottom of the frame [Fig. T7]. Create copy of the cube and then align it as shown in Fig. T7.

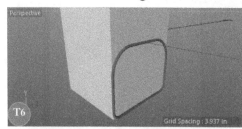
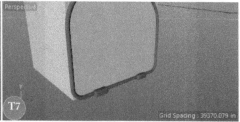

6. Select everything in **Object Manager** except **Cube** and then press **Alt+G** to group the objects. Rename the group as **Frame.1**. Create a copy of the **Frame.1** and align it with the other side of the **Cube** [Fig. T8]. Hide **Cube**.

⊙→ *What next?*
Now, we will create the seat and back support.

7. Create a plane and then in **Attribute Manager**, set **Width, Height, Width Segments**, and **Height Segments** to **23, 20, 1,** and **1,** respectively. Align the plane with the frame [Fig. T9].

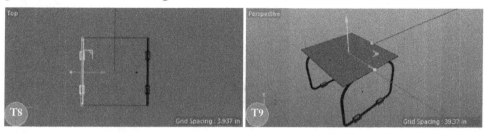

8. Make the plane editable by pressing **C**. Select the edge [Fig. T10] and then drag it upward about **15** units with the **Ctrl** held down to create the back support [Fig. T10]. Similarly, extrude the front edge using **Ctrl** [Fig. T11].

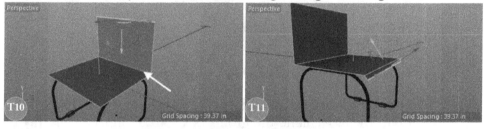

9. Select the edge [Fig. T12] and then bevel it using the **Bevel** tool [Fig. T13].

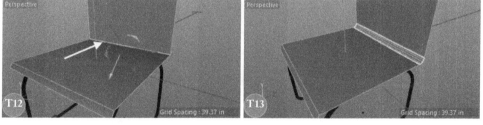

10. Use the value **1** for **Offset** and **3** for **Subdivision** in the **Attribute Manager | Bevel | Tool Option** tab. Create edge loops using the **Loop/Path Cut** tool [Fig. T14].

11. Select the edge ring, as shown in Fig. T15 and then press **MM** three times to connect the edges [Fig. T16]. Select all polygons of the **Plane** and then extrude them by **0.169** units using the **Extrude** tool [Fig. T17].

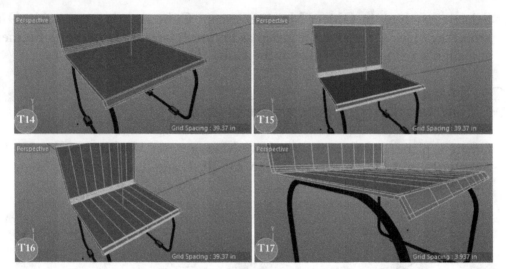

12. Now, you need to create a bend in the seat and back support using the **FFD** deformer. Finally, add a **Subdivision Surface** generator to smooth the chair [Fig. T18].

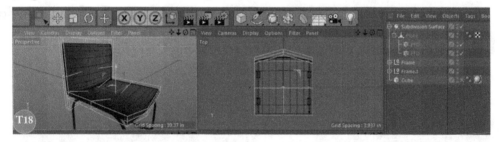

13. Delete **Cube** from the scene. In **Object Manager,** select all objects and press **Alt+G** to group them. Rename the group as **Chair.** Connect an **Array** object with **Chair.** In the **Attribute Manager | Array Object | Object** tab, set **Radius** and **Copies** parameters as required.

Appendix AM: Quiz Answers

Multiple Choice
1. B, 2. B, 3. A, 4. C, 5. B

Fill in the Blanks
1. lock, unlock, 2. C, 3. Shift+S, 4. Ctrl+D, 5. MMB, 6. Shift+C, 7. Shift+F4, 8. Shift+G

True/False
1. T, 2. T, 3. T, 4. T, 5. F

Chapter M2: Tools of the Trade

Multiple Choice
1. A, 2. A

Fill in the Blanks
1. Workplanes, 2. Center, 3. Transfer, 4. Lens Distortion, 5. Doodle

True/False
1. F, 2. T, 3. T, 4. T, 5. F

Chapter M3: Spline Modeling

Multiple Choice
1. D, 2. C, 3. A, 4. C, 5. D

Fill in the Blanks
1. C, 2. Spine Arch Tool, 3. Intermediate Points

True/False
1. T, 2. T

Multiple Choice
1. A, 2. B, 3. B, 4. A

Fill in the Blanks
1. Points, Edges, and Polygons, 2. C, 3. Set Selection

True/False
1. T, 2. F

Index

I

T